The U.S. Microelectronics Industry

The Technology Policy and Economic Growth Series
Herbert I. Fusfeld and Richard R. Nelson, Editors

Fusfeld/Haklisch INDUSTRIAL PRODUCTIVITY AND INTERNATIONAL TECHNICAL COOPERATION
Fusfeld/Langlois UNDERSTANDING R&D PRODUCTIVITY
Hazewindus THE U.S. MICROELECTRONICS INDUSTRY: Technical Change, Industry Growth and Social Impact
Nelson GOVERNMENT AND TECHNICAL PROGRESS: A Cross-Industry Analysis

Pergamon Titles of Related Interest

Dewar INDUSTRY VITALIZATION: Toward a National Industrial Policy
Hill/Utterback TECHNOLOGICAL INNOVATION FOR A DYNAMIC ECONOMY
Lundstedt/Colglazier MANAGING INNOVATION: The Social Dimensions of Creativity
Perlmutter/Sagafi-nejad INTERNATIONAL TECHNOLOGY TRANSFER
Sagafi-nejad/Belfield TRANSNATIONAL CORPORATIONS TECHNOLOGY TRANSFERS AND DEVELOPMENT
Sagafi-nejad/Moxon/Perlmutter CONTROLLING INTERNATIONAL TECHNOLOGY TRANSFER

Related Journals*

BULLETIN OF SCIENCE, TECHNOLOGY AND SOCIETY
COMPUTERS AND INDUSTRIAL ENGINEERING
COMPUTERS AND OPERATIONS RESEARCH
SOCIO-ECONOMIC PLANNING SCIENCES
TECHNOLOGY IN SOCIETY
WORK IN AMERICA INSTITUTE STUDIES IN PRODUCTIVITY

*Free specimen copies available upon request.

The U.S. Microelectronics Industry

Technical Change, Industry Growth and Social Impact

Nico Hazewindus
with John Tooker

The Technology Policy and Economic Growth Series,
Herbert I. Fusfeld and Richard R. Nelson, Editors

Published in cooperation with the Center for Science and Technology Policy,
Graduate School of Business Administration, New York University

Pergamon Press
New York Oxford Toronto Sydney Paris Frankfurt

Pergamon Press Offices:

U.S.A.	Pergamon Press Inc., Maxwell House, Fairview Park Elmsford, New York 10523, U.S.A.
U.K.	Pergamon Press Ltd., Headington Hill Hall, Oxford OX3 0BW, England
CANADA	Pergamon Press Canada Ltd., Suite 104, 150 Consumers Road, Willowdale, Ontario M2J 1P9, Canada
AUSTRALIA	Pergamon Press (Aust.) Pty. Ltd., P.O. Box 544, Potts Point, NSW 2011, Australia
FRANCE	Pergamon Press SARL, 24 rue des Ecoles, 75240 Paris, Cedex 05, France
FEDERAL REPUBLIC OF GERMANY	Pergamon Press GmbH, Hammerweg 6 6242 Kronberg/Taunus, Federal Republic of Germany

Copyright © 1982 Pergamon Press Inc.

Library of Congress Cataloging in Publication Data

Hazewindus, Nico, 1937-
  The U.S. microelectronics industry.

  (The Technology policy and economic growth series)
  Includes index.
  1. Microelectronics industry--United States.
2. Microelectronics industry--United States--
Technological innovations. 3. Microelectronics
industry--Government policy--United States.
I. Tooker, John. II. Title. III. Series.
HD9696.A3H556   1982   338.4'76213817'0973   82-12191
ISBN 0-08-029376-X

All Rights reserved. No part of this publication may be reproduced, stored in a retrieval system or transmitted in any form or by any means: electronic, electrostatic, magnetic tape, mechanical, photocopying, recording or otherwise, without permission in writing from the publishers.

Printed in the United States of America

Contents

PREFACE - NICO HAZEWINDUS AND JOHN TOOKER		ix
ACKNOWLEDGMENTS		xi
GLOSSARY		xiii

CHAPTER

1   A PERSPECTIVE ON MICROELECTRONICS   1

       Theme and Scope   1
       Microelectronics - an Introduction   3
       Terminology   5

2   INTEGRATED CIRCUITS AND THEIR IMPACT: THREE EXAMPLES   8

       Characteristics of Integrated-circuit Technology   9
       Computers   16
       Telecommunications   19
       Control Systems   23

3   MICROELECTRONICS: THREE APPLICATION AREAS   27

       Consumer Electronics   27
       The Office   32
       Factory Automation   37
       Assessment of the U.S. Position   39

CHAPTER

4	INTEGRATED CIRCUITS, TECHNOLOGY	44

    IC Technology 46
    Transistors and Integrated Circuits 46
    Basic Manufacturing Technology 51
    IC Design and Manufacturing 53
    Current Status 56

5	INTEGRATED CIRCUITS, PRODUCTS	59

    Digital Memories 59
    Microprocessors 65
    General and Specific Digital Logic 69
    Uncommitted Logic Arrays 70
    Analog IC's 71
    Micellaneous Devices 74
    Some Final Notes 77

6	THE INTEGRATED CIRCUIT INDUSTRY	79

    IC'S: A Worldwide Business 80
    The U.S. Merchant Industry 82
    The Captive Industry 89
    Support Industry 96
    The European Industry 97
    The Japanese Industry 100

7	TRENDS: A LOOK AT THE FUTURE OF THE IC INDUSTRY	107

    Developments in VLSI Technology 108
    VLSI Design 110
    Several Challenges 115
    Structural Development of the U.S. Industry 116
    New Ventures and Innovation 119
    Government Support Strategies 120

8	THE TECHNOLOGY BASE OF THE U.S. INTEGRATED CIRCUIT INDUSTRY	129

    The Role of Academic Research 130
    University Research Programs 131
    Industry-University Relations 138
    The Federal Government 141
    Industrial R&D 147

CONTENTS vii

CHAPTER

9 THE MANPOWER PROBLEM 151

   Data 151
   Consequences for Education 154

10 MICROELECTRONICS AND EMPLOYMENT 158

    A Survey of the Literature 159
    Governments and Labor Unions 165

11 POLICIES OF FEDERAL AND STATE
    GOVERNMENTS 169

    Attitudes of the Federal Government 171
    Department of Commerce 173
    State Department and Justice
     Department 174
    National Science Foundation 175
    Department of Defense 177
    The SIA and Public Policy 178
    Policies of the States 180
    Concluding Remarks 180

12 U.S. MICROELECTRONICS 184

    Main Themes: A Recapitulation 185
    Actions and Policies 188
    Final Remarks 191

INDEX 195

ABOUT THE AUTHORS 199

Preface

This book is intended to serve as a guide to the interactions of technical change, industry growth, and public policy in the field of microelectronics. The approach is a broad technical perspective on the next five to ten years. It provides an introduction to the technology of microelectronics, the structure of the industry, the base of research and manpower supporting this capability, and selected government policies which can affect the vitality of this exceptional field.

This study thus presents an integrated view of the diverse facets that characterize microelectronics. It can serve as a reference for considering, in greater depth, areas of special concern to policymakers in both the public and private sectors who require a sense of the whole system to understand better their own particular interests.

This book was conceived and written during the author's half-year stay in 1981 as a Visiting Research Fellow at the Center for Science and Technology Policy, Graduate School of Business Administration, New York University. This stay marked a leave of absence from the Product Development Coordination Bureau of N.V. Philips' Gloeilampenfabrieken in The Netherlands. The material contained here is based on discussions during that time with experts in government, industry, and universities whose knowledge and experience brought special insight to the research. A diverse range of references is included in this book and may serve as a substantive base of background information. To this the author added his own opinions and conclusions and the final text therefore remains his sole responsibility.

Eindhoven/New York, March 1982
Nico Hazewindus with the
collaboration of John Tooker

Acknowledgments

I would like to thank Dr. A.E. Pannenborg, Vice President of the Board of Management, N.V. Philips, who initiated my stay at the Center for Science and Technology Policy of New York University. I am indebted to Dr. Herbert I. Fusfeld, Director of the Center, for his guidance and assistance during this period. I thank Dr. P. Kramer, Director of the Product Development Coordination Bureau at Philips, for his support during my stay at the Center as well as during the period of preparation of this manuscript.

I gratefully acknowledge the cooperation of my colleagues at the Center. Dr. Lois S. Peters introduced me to the topic of university-industry relations. Mrs. Carmela S. Haklisch critically read several versions of the manuscript and suggested many changes. Her efforts have contributed substantially to the completion of the book.

The material contained here has been strongly influenced by discussions with members of an ad hoc advisory group. The participants of this group included Dr. Erich Bloch, Vice President, Technical Personnel Development, IBM; Dr. W.F. Brinkman, Director, Chemical Physics Research Laboratory, Bell Laboratories; Dr. W. Chu, Manager, Technology Development Operation, G.E.; Dr. David Golibersuch, Manager, Signal Electronics Lab, G.E.; Dr. Donald King, President, Philips Laboratories, North American Philips Corporation; Dr. Henry Kressel, Staff Vice President, Solid State Research, RCA; Mr. Joseph Reed, Technical Director of Military Communications, ITT Corporation; Dr. Frank A. Sewell, Jr., Director, Semiconductor Laboratory, Sperry; and Dr. Lee L. Davenport, Vice President and Chief Scientist (retired), GTE. Dr. Davenport's continuing interest assisted me throughout the study in various discussions, and I thank him for his very helpful review of the first draft of this manuscript.

ACKNOWLEDGMENTS

I am grateful to many people in the U.S. and Europe who helped me, either by sharing their ideas in interviews or by sending me materials and information.

At the Center for Science and Technology Policy my collaborator John Tooker was the research assistant for the project. Among his many contributions I mention especially his work on the sections on employment effects, the IC industry, various national policies, and manpower needs.

Maria Ortiz and Shawn Roberts patiently typed the manuscript and Carlos Santiago drafted all the illustrations.

Finally, I would like to thank everyone at the Center for their gracious hospitality which made my stay such a memorable occasion.

Nico Hazewindus
Eindhoven, March 1982

Glossary

Actuator	–	Electronically driven device that affects some mechanical movement.
Active component	–	Electronic component like a diode or transistor used to switch or amplify electric signals.
ADA	–	High-level computer language developed for the U.S. Department of Defense.
A/D converter	–	Electronic circuit that converts analog signals into a digital representation.
Algorithm	–	Rule or procedure for solving a mathematical problem.
Analog	–	Electric signal with continuously varying amplitude (usually used as opposed to digital).
Application software	–	Software that is needed to apply a general computer to a specific task.
Arpanet	–	U.S. data network linking computers nationwide.
BASIC	–	Relatively simple high-level computer language.
Bipolar device	–	Transistor device consisting of two types (n and p) semiconductor.

GLOSSARY

Bit	-	Unit of information, consisting of a binary digit having the value 0 or 1.
Bit-rate reduction	-	Technique to diminish the number of bits per second needed to represent sound or moving pictures by eliminating redundancy.
Bubble memory	-	Device to store data in magnetized spots in a thin layer of magnetic material.
Bus	-	Electronic circuit used to provide a (standardized) path for data exchange between devices.
CAD	-	Computer-Aided Design, a method to design electronic circuits or mechanical parts with the help of a computer.
CAM	-	Computer-Aided Manufacturing, the application of computers in several stages of the industrial production process.
Captive producer	-	Company producing IC's for its own use, in contrast with merchant houses.
CCD	-	Charge-Coupled Device, a semiconductor device in which information is stored as small electronic charges.
Chip	-	Tiny piece of semiconductor crystal on which an integrated circuit has been made.
CIF	-	Caltech Intermediate Format, a computer language to describe the layout of integrated circuits.
CMOS	-	Complementary MOS, integrated circuit with two types (n and p) MOS transistors, which uses little power.
COBOL	-	High-level computer language for business applications.
CODEC	-	Coder-decoder circuit used in telephony to convert analog signals to digital ones and vice versa.

GLOSSARY

D/A converter	-	Electronic circuit to convert digital signals into analog ones.
DARPA	-	Defense Advanced Research Projects Agency (U.S.).
Database	-	Large collection of digital data, arranged in a predetermined way and stored in a computer memory.
DES	-	Data Encryption Standard, method for encryption of digital data, devised by IBM and accepted by the U.S. government for certain classes of application.
Die	-	Tiny piece of semiconductor crystal on which an integrated circuit is made.
Digital	-	Representation of a signal as a number, as opposed to analog.
Diode	-	Active electronic device conducting current in one direction, but not in the other.
DMOS	-	Double-diffusion MOS, devices which are suitable for high-voltage or high-power applications.
Doping	-	Introduction of impurities into a semiconductor to influence its electric characteristics.
Dynamic RAM	-	Semiconductor random access memory requiring a continuous refreshment of stored data.
E-beam machine	-	Equipment using a high-energy electron beam to make very fine patterns in the lithography for integrated circuits.
ECL	-	Emitter Coupled Logic, class of bipolar integrated circuits.
EDP	-	Electronic data processing.
EE PROM	-	Electrically Erasable PROM, an E PROM that can be erased with electric pulses.

GLOSSARY

E PROM	-	Erasable Programmable Read-Only Memory, PROM of which the contents can be erased with ultraviolet light.
Epitaxial layer	-	Thin layer, deposited on a semiconductor material, with a similar crystal structure.
FET	-	Field Effect Transistor, a transistor based on conduction effects, controlled by a voltage, in one kind of semiconductor material (also called unipolar transistor).
Fibers (optical)	-	Thin transparent glass fibers which transmit information by means of modulated light.
Firmware	-	Software "packaged" in hardware form.
Floppy disc	-	Flexible magnetic disc used for storage of data in a computer.
FORTRAN	-	High-level computer language, in particular suited for mathematical calculations.
Gallium Arsenide	-	Semiconductor material, in particular used for high-speed transistors.
Gate	-	Basic logic circuit element that determines whether certain logic conditions at its inputs are simultaneously met (e.g. "and", "or" gates).
Gate arrays	-	Integrated circuits containing an array of gates, which are connected by means of a conductor pattern to perform a certain function.
GATT	-	General Agreement on Tariffs and Trade.
Hall sensor	-	Sensor making use of the Hall effect, converting a magnetic field into an electric signal.
Hardware	-	The various physical parts of a computer system, contrasted with software.

GLOSSARY

Hybrid circuits	– Combinations of active and passive components, mounted on a substrate (usually ceramic) on which resistors and connecting conductors are applied.
IC	– Integrated Circuit, semiconductor device incorporating a number of active and passive circuit elements on a single piece of semiconducting material (chip).
IIL, I^2L	– Integrated Injection Logic, bipolar integrated circuit with a particular arrangement of diodes and transistors to achieve good packing density at reasonable speed.
Interface	– Place where two equipments must be matched in order to be connected; also used as "man-machine interface"
I/O	– Input/Output, usually designating these functions in a computer system.
Ion implantation	– A method of introducing impurities in a semiconducting material by bombarding the latter with high-energy ions.
ISL	– Integrated Schottky Logic, particular type of bipolar integrated circuit with properties like IIL.
ITAR	– International Traffic in Arms Regulations (U.S.)
Josephson junction	– Superconducting semiconductor device operating at cryogenic (near absolute zero) temperatures, with extremely fast operation speed.
Junction	– Boundary between two types of semiconducting material where physical phenomena occur that are utilized in diodes or transistors.
Learning curve	– The graph showing a lowering of manufacturing costs with a certain factor every time the total series produced is doubled.
Linear IC	– Analog IC, as opposed to digital IC.

GLOSSARY

Lithography	-	Processes to make patterns needed in integrated circuits, using light, X-rays, or electron beam to define the patterns and etching techniques to remove undesired parts.
LSI	-	Large-scale integration, IC's containing 1,000 to 100,000 components.
Mainframe	-	Medium or large computer without peripherals.
Masks	-	Glass plates with IC layouts, used in the lithographic process.
Memory	-	Store for digital or analog information, kept usually in electric or magnetic form.
Merchant house	-	Company producing IC's to sell them on the open market.
Microelectronics	-	The technology and manufacture of miniature electronic components and circuits.
Micron	-	One thousandth of a millimeter.
Microprocessor	-	Central part of a computer integrated on an IC.
Minicomputer	-	Small computer, usually much less expensive than a mainframe.
MITI	-	Ministry of International Trade and Industry (Japan).
Modem	-	Modulator-Demodulator, circuit used to couple digital equipment to an (analog) telephone line.
MOS	-	Metal-Oxide-Silicon, technology to produce transistors controlled by voltage (MOS FET).
Moore's Law	-	Statement that complexity of IC's increases annually with a certain factor.

GLOSSARY

MSI	-	Medium Scale Integration, IC's containing 100 to 1,000 components.
Nano second	-	One thousandth of a millisecond.
NBS	-	National Bureau of Standards (U.S.)
NIH	-	National Institutes of Health (U.S.)
nMOS	-	MOS transistor using electrons to conduct the current.
Non-volatile memory	-	Memory of which the content is not destroyed when electric power is switched off.
NSF	-	National Science Foundation (U.S.)
n-silicon	-	Silicon doped to have an excess of electrons.
OECD	-	Organization for Economic Cooperation and Development.
Opto-electronics device	-	Semiconductor device used in optical systems, for instance to generate or detect light.
Operational amplifier	-	Analog amplifier for general application, the properties of which can be determined by the circuit designer.
PABX	-	Private automatic branch telephone exchange, telephone switch for private (business) networks.
PASCAL	-	High-level computer language.
Passivation layer	-	The last, protecting layer that is applied on an integrated circuit.
Passive components	-	Electronic components like resistors and capacitors which do not actively influence signals in a circuit.
Piezo-electricity	-	Physical phenomenon in certain materials in which mechanical stress is transformed into electricity, and vice versa.

GLOSSARY

Printed circuit board — Electrically isolating board with a pattern of conductors, on which electronic components are soldered.

Programmable logic array — Circuit with a prearranged array of logic gates which can be interconnected to implement a desired logic function.

Programming language — A coherent set of instructions resembling (English) language that can be understood by a computer.

PROM — Programmable Read-Only Memory, a memory in which the user can write data once. Thereafter only read operations are possible.

p-silicon — Silicon doped to have a shortage of electrons.

PTT — Governmental organization which is, in many countries, responsible for mail and telecommunications.

RAM — Random Access Memory, memory in which any part of the stored information can directly be accessed, read, and modified.

Resist — Light-sensitive lacquer used in the lithographic process of IC manufacturing.

Reticle — Glass plate with pattern to be projected on a silicon wafer in IC lithography.

ROM — Read-only memory, a memory of which the content has been determined during manufacturing; thereafter only reading operations are possible.

SAW device — Surface Acoustic Wave device, used in electronic filters or resonators, based on the phenomenon that different types of acoustic waves may travel on the surface of certain materials.

Second sourcing — Practice among merchant houses to allow the manufacturing of a particular proprietary IC by a competitor, thus giv-

GLOSSARY   xxi

	ing one's customer the certainty of having two independent suppliers.
Schottky effect	- Particular effect on semiconductor material surfaces that can be used to make diodes.
Sensor	- Device that transforms a physical phenomenon into a measurable (often electric) signal.
Semiconductor	- Device employing a material (or the material itself) with electrical characteristics in between a conductor and an isolator.
SIA	- Semiconductor Industry Association, U.S. trade association.
Silicon Valley	- Region south of San Francisco (California) where many IC companies have their headquarters.
SLIC	- Subscriber Line Interface, circuit to interface a digital telephone exchange and analog subscriber lines.
Slice	- Thin wafer of silicon on which IC's are manufactured.
Software	- Programs used to instruct a computer.
SOS	- Silicon-on-sapphire, technique for making very fast transistors on a sapphire substrate.
SSI	- Small-scale integration, less than 100 components per IC.
Static RAM	- Random-access memory of which the content need not be refreshed like a dynamic RAM.
Systems house	- Electronics firm having its business in electronic systems.
Teletext	- TV-broadcast format in which pages of digital information are transmitted simultaneously with the normal video signals.

GLOSSARY

TDM	-	Time Division Multiplex, method to combine different transmissions of digital information into one signal.
TFT	-	Thin-film transistor, transistor array made in a thin film of amorphous semiconducting material.
Thick film circuit	-	Technique in which conductors and resistors are screen-printed on an (alumina) substrate.
Thin film circuit	-	Technique in which conductors and resistors are lithographically applied on a thin substrate.
Transistor	-	Active semiconductor device used as amplifier or as a switch.
TTL	-	Transistor-Transistor Logic, class of bipolar digital IC's with a certain configuration of diodes and transistors.
Uncommitted Logic Array	-	Array of logic gates in the form of an IC which can be connected in a last metallization layer (synonym for gate array).
VHSIC program	-	Very High Speed Integrated Circuit program of the U.S. Department of Defense, designed to provide high-speed signal processors for military purposes.
Viewdata	-	System to connect a user with a TV set to a central computer though a telephone line.
VLSI	-	Very Large Scale Integration, more than 100,000 components per IC.
VMOS	-	MOS technology in which V-shaped grooves are used to isolate transistors.
Wafer	-	Slice of semiconductor material (usually silicon) on which IC's are manufactured.
Word	-	Number of bits processed by a computer or other digital device as one unit.

GLOSSARY

X-ray lithography - Lithographic process using X-rays instead of visible light to arrive at smaller dimensions.

ZMOS - Type of MOS technology usually applied in high-power devices.

1
A Perspective on Microelectronics

THEME AND SCOPE

Technical change has always been a driving force influencing the history of mankind. The printing press, navigational aids, and the steam engine profoundly altered the lives of our ancestors. Thus, in an historical perspective, the emergence of microelectronics and its effects on our society may seem to follow a predictable, or at least an understandable, course.

There are, however, two main reasons why microelectronics does not fit in an historical analogy. First, there is the tremendous pace of its development and application. In a few decades a complete technical revolution has occurred and continues to unfold in the world of electronics. Second, the technology has an extremely pervasive character affecting virtually all sectors of our society. It has an unparalleled influence on our quality of life, patterns of professional activities, and economic, political, and international relationships.

Thus, microelectronics presents a unique case of the profound and far-reaching effects that technical change can have. We are now midway in the process of response to this change: it began roughly a decade ago and it will most likely continue through the remainder of this century. This implies that though major impacts have already become evident, many have only barely surfaced or are still hardly recognizable.

To understand the effects of this technical change, one must understand its main driving force: microelectronics technology. This technology is itself still in a continuous state of change, primarily caused by rapid developments in the field of integrated circuits which can be considered the "building blocks" of microelectronics.

However, the unique characteristics of the technical change brought about by microelectronics and the complexity of its impacts have consequences that go far beyond an explanation based solely on an understanding of the technology. Of critical importance is an understanding of the context in which this technical change is taking place, in particular the factors forming the major response to the changes that microelectronics is bringing about. This response thus requires a perspective that integrates the driving force of the technology with the tapestry of relationships among industry, government, academia, and society.

Many books and articles have discussed particular aspects of microelectronics; and yet a coherent overview of this field, influenced as it is by science and technology, industry and trade, federal policy and public acceptance, has not received much attention. This book is an attempt to provide such a perspective, giving one an understanding of the subject as a whole and insight into how the different parts are related. It should therefore be relevant to anyone whose interests or professional responsibilities relate to the field - scientists, engineers, investors, research managers, industry executives, university faculties, and government policymakers. Each of these individuals may be involved with some dimension of microelectronics, and each must often make decisions within a specialized area of expertise for which a balanced view of other key issues in the field is needed.

This book thus offers basic information that can be instrumental in analyzing the complicated issues associated with microelectronics. It is an overview of the terrain and an approach to understanding the intricate relationships that exist. The book includes four main areas: (1) a discussion of integrated circuits (IC's) as the cornerstone of a multitude of applications in numerous types of electronic equipment, (2) a description of the production technology and main products of IC's and an overview of the organization of the industry, (3) an examination of the research and manpower base supporting technology advances, and (4) a review of government policies and their relationship to technical change and industry vitality.

The issues covered here are viewed in a perspective that is basically technical and prognostic, with the focus on the U.S., the country where many of the relevant technologies have originated and made their greatest impact. Many other perspectives provide equally interesting viewpoints. Studies have been published dealing, for instance, with international trade, international competitive environments, or the history of technology transfer. The approach used here has the advantage that it fully recognizes the "technology push" that is so evident in microelectronics. It underlines the belief that the impact of microelectronics has just begun, and that the real pervasiveness of the technology and the related response will be unveiled over the next decade.

One might argue that prospective views for the next ten years are by necessity inaccurate; new technologies may even completely reverse current trends and projected impacts. Nevertheless, given the present situation, a forceful development of current technologies is likely to continue with a wide field of application for the ensuing products. This combination of technology push and market pull suggests that considerable effort will be spent on the development of microelectronics along fairly predictable lines. Any other new technologies will require large investments in R&D and further industrialization before becoming major competitors. Hence the likelihood that such new developments will seriously influence the present five - to ten - year outlook is relatively remote.

In summary, this book should provide the reader with a better understanding of the complexity characterizing the field of microelectronics. In this way, it may contribute to the insight of anyone who has, in whatever capacity, interests, or responsibilities related to this remarkable technology.

MICROELECTRONICS - AN INTRODUCTION

A few years after World War II, microelectronics was born in a solid-state physics research group at Bell Laboratories. The invention of the transistor started off a chain of scientific, technological, and industrial developments which ultimately have established microelectronics as the most important technological driving force of modern society.

Electronic equipment, based on transistors and later on integrated circuits, has shown a tremendous growth in performance, combined with, surprisingly enough, a sharp reduction in the costs. A dramatic example is the pocket calculator, where performance increase and cost reduction give present-day elementary-school children a computing power that their parents might not have found in a university computer center.

These characteristics of microelectronics, notably more performance at less cost, have opened up a vast array of applications, thereby replacing older technologies in existing technical functions or adding completely new functions to the existing spectrum. Examples can be found readily in computers, telecommunications, consumer electronics, manufacturing technology, and office systems, to name just a few.

The U.S. has been particularly adept in developing and exploiting microelectronics technology. This has been done largely through business enterprises. Several large, long-established electronics firms have successfully transformed their activities into the microelectronics era, but many others have not. A distinct trend has been the emergence of new ventures: small companies developing and exploiting some form of advanced technology. Several of those have grown explosively and have become leaders in their field, often overshadowing traditional powers in the marketplace.

Thus, technical change has brought about tremendous opportunities for industrial growth, which some companies have capitalized upon to prosper, while others, not seizing opportunities, have faded away. Following the first wave of entrepreneurial activity in the last two decades, developments expected in the near future are less clear. It seems that the present distinction between component-oriented companies and system-oriented firms has begun to diminish. This trend can be explained both by the current direction of technological change and by economic considerations. Opportunities for new ventures still exist, though perhaps less abundantly than before: a highly selective process is very likely the trend of the future.

International competition is a factor increasingly worrying the U.S. industry. Japan in particular has become an important competitor in recent years. Backed by coordinating actions of their government, Japanese manufacturers have captured large shares of some important markets. Concerted actions for further industrial growth in Japan, based on technological excellence, have been initiated, inspiring awe in U.S. governmental and industrial circles. This will also influence the structure and relations of industry in this country.

A high-technology industrial activity like microelectronics is embedded in U.S. society in many different ways. For instance, industrial progress depends strongly on universities and technical schools for basic scientific and technical knowledge as well as for education of skilled scientists and engineers. Many academic institutions, however, have not been able to respond adequately to the rapid pace of technological change, either in training or research. This situation is being remedied at several institutions where initiatives, both from academia and from industry, have been taken that may lead to new efforts, often cooperative, at the frontiers of science and engineering. This development will influence relations between academia and industry as they have grown in the past decades.

A more general issue is the impact that microelectronics is supposed to have on the structure of society. A major one relates to changes in employment brought about by the new technology. It is evident that many jobs will be eliminated, for instance resulting from automation in offices and factories. But new technology also provides opportunities for new services and new products, thus creating many new jobs. This implies that the jobs of many people will change, that measures are required to support those affected by automation, and that they and others must be trained in the newly needed skills.

At the present time no realistic insights exist, qualitatively or quantitatively, into these effects on a national scale. Much will depend on how microelectronics will actually be used. Though the spectrum of possible applications is staggering, a relatively limited selection of uses will find widespread acceptance. Any prediction here must first be based on assessments of present and future technologies. Thereafter, other factors like usefulness, economic advantage, human acceptability, and the necessary capital investment will strongly determine the future of such technological innovations.

Microelectronics has influenced - and will continue to influence - U.S. society in many important ways. Not surprisingly, the Federal government is increasingly confronted with the need for public policies in microelectronics-related areas. It has to secure a first-rate technology base for reasons of national security. It has to consider adverse impacts on such a key industry resulting from economic policies and international competition. It has to regulate new applications of microelectronics to secure fair opportunities to manufacturers and users. It may be forced to react to employment problems as they emerge.

It is obvious that public policy in this field is an extremely complex matter. It should be tuned to developments in the field of microelectronics, to the technical and economic status of the industry and its structural changes, and to the concomitant impacts on society resulting from the introduction of new technologies.

TERMINOLOGY

Before embarking on detailed discussions of microelectronics, it may be helpful to the reader to provide a brief introduction to the terminology used. Electronics emerged in the 1920's as a separate discipline in engineering when suitable passive components (resistors, capacitors) and active components (electron tubes) became available. They were mounted on an aluminum chassis and wired together into circuits that performed a certain function. A number of such circuits put together into a package finally formed the apparatus.

In the 1950's semiconductor devices like diodes and transistors replaced most of the electron tubes in low-power circuits. The use of these discrete semiconductor devices led to an appreciable reduction in the volume and the power-consumption of the circuits. At the same time, new mounting techniques like printed-circuit boards were developed. The result was a new generation of low-cost, long-life circuits which found many applications.

Integrated circuits (IC's), gaining use in the 1960's and 1970's, combined several transistors and their interconnections on one piece of semiconducting material, a chip. In this way, a function could be encased in one small, low-power device. Several IC's together, often mounted on a printed-circuit board (though new techniques using ceramic substrates also came into use), formed the basis of the apparatus.

The number of transistors that can be integrated together on one chip has greatly increased over the years. Terms such as small-scale integration (SSI), medium-scale integration (MSI), large-scale integration (LSI), and very large-scale integration (VLSI) were coined to indicate the increasing density. Currently, the era of VLSI in which very complex functions can be accomplished on one chip is slowly coming into being.

Microelectronics is concerned with the design and application of such high-density integrated circuits. In the early years of electronics, design and manufacturing of components, tubes, transistors, and resistors could be completely separated from the circuit design. Traditionally, the designer used a little theory and a lot of practical experimentation to arrive at the proper form of circuits. Now, however, the designer's task is much more complex. Design architecture has become important, the use of software tools for design and simulation has been introduced, software programming has been required for many designs, and integrated circuits are often not only purchased but also designed. This increased task for the circuit designer has required substantial development of suitable new techniques and tools, and relatedly the education and training of individuals with the requisite new skills.

Microelectronics has enabled the production of equipment that is small and cheap but at the same time has a high functional complexity. This has led to many new types of electronics equipment, either because they perform a new function at an affordable price or because they yield cost-effective solutions compared with other technologies or with human labor. Here lie the major causes for the pervasiveness of microelectronics in the industrialized societies.

The largest part of integrated-circuit production is in standard types, offered as general-purpose IC's targeted for use by many customers. Adaptation to the needs of a specific customer can be accomplished by adding external circuitry to the IC or, if possible, by programming it. In many cases, the user wants more specific, specially designed IC's, leading to the existence of a custom-IC activity.

PERSPECTIVE ON MICROELECTRONICS

The industrial activity of providing IC's for the market is largely in the hands of the "merchant industry" companies which produce IC's for a wide range of customers. In contrast, many systems companies have decided to establish their own in-house manufacturing facility. Such "captive activities" have been increasing in recent years. Many IC manufacturers in Europe and Japan have both a merchant program and captive activities.

2
Integrated Circuits and Their Impact: Three Examples

Technical change has often resulted in new developments in society. Microelectronics is perhaps the ultimate example of this. In an industrial society, already greatly dependent on technologies like computers and communication systems, microelectronics has provided the technological base for an "information society." This development has gained considerable momentum in the past decade,(1) but it seems certain that still more spectacular technology-induced changes will occur.

What is going to happen will largely depend on today's technology in a few key sectors. In this chapter, we shall review them and also suggest several lines of future developments.

First, however, we have to briefly consider integrated circuits (IC's) which have had a tremendous impact on the technologies we will discuss. What are the major properties of IC's that make them such a useful "building material" for the electronics industry? Some elements of an answer to this question can be found by taking a closer look at a particular circuit, the Metal Oxide Semiconductor (MOS) Random Access Memory, generally considered to be a premier driver of the development of IC technology.

After this description of the general properties of IC's, their influence will be considered on three important sectors of the electronic industry: computers, telecommunications, and control systems. These techniques form the basis for the application of microelectronics in many other fields.

INTEGRATED CIRCUITS AND IMPACT

CHARACTERISTICS OF INTEGRATED-CIRCUIT TECHNOLOGY

Though integrated circuits (IC's) will be described in much detail in later chapters, some important characteristics may be understood from a discussion of a specific example. This example is a particular memory circuit, the random-access memory (RAM) that is based on a certain semiconductor technology, called nMOS. Generally this type of circuit is seen as the cutting edge of the technology. Fierce international competition has affected the performance and price of this circuit perhaps somewhat differently than is true for the typical IC. Nevertheless, it provides a good vehicle for this discussion.

The Memory Circuit

Electronic circuits perform many operations, ranging from amplification of analog signals to the multiplication of two numbers which are given in a binary form (i.e., "zeros" and "ones"). Integrated circuits consist mostly of a large number of diodes, transistors, resistors, and capacitors manufactured on the surface of a tiny silicon crystal, which are interconnected via conducting paths that run across this surface.

A digital memory(2) consists of an array of cells, each consisting of several transistors. Each cell can store one unit (a "bit") of binary information, a "zero" or a "one." In a random-access memory, every cell can be addressed separately in order to read the existing content or to write a new content.

In the particular nMOS technology, each cell consists of only one transistor and a small capacitor. The presence or absence of a charge on the capacitor determines whether a "one" is stored or a "zero." This has proved to be a very efficient design in terms of the silicon area needed per cell. Consequently, it appears possible to pack large numbers of cells in one integrated circuit.

Moore's Law

The number of components per integrated circuit has doubled every year in the last two decades, as illustrated in figure 2.1.(3)

Generally speaking, this phenomenon is due to the tremendous advances in technology which the industry has achieved. Cunningham(4) provides a useful analysis for this discussion.

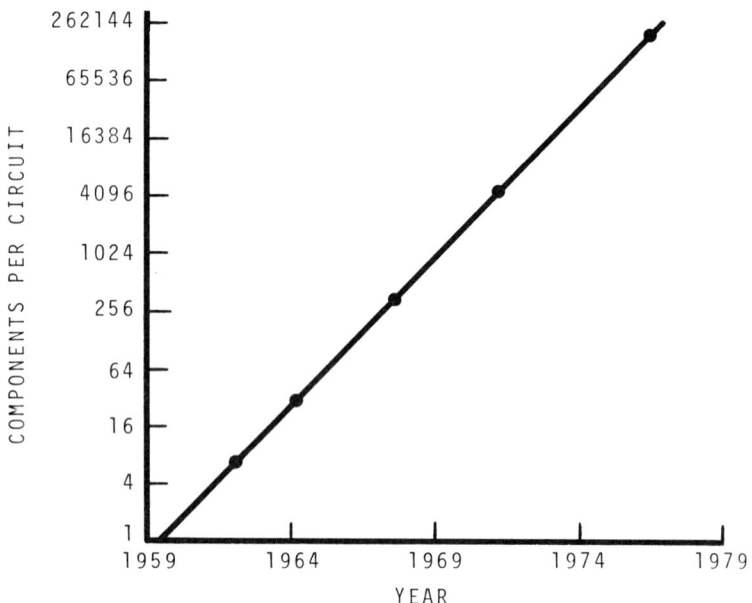

Fig. 2.1. Moore's law.

In the mid-1960s it was possible to construct a 64-bit RAM. In 1981, 16k bit (i.e., 16,000 bit) RAM's were routinely available (64k-bit RAM's are entering the market and 256k-bit RAM's have been demonstrated in the laboratory). This increase of a factor of 256 in components per chip was attained in the following way. (See also Fig. 2.2.)

- Diminishing the cell area accounts for a factor of 16 increase in memory capacity. This has entailed inventing new cell structures and different layouts and reducing the size of transistors and capacitors, made possible by a strongly increased manufacturing process control.
- The chip size could be enlarged, to give an improvement of a factor of 4. Better understanding and control of the silicon base material and enhanced process technologies are mainly responsible.
- The design rules (the line widths used in the actual circuit - e.g., conductors) could be shrunk: this accounts again for a factor-4 improvement. Here, better optical equipment with increased resolution was needed, more discriminate etching techniques were developed, and so forth.

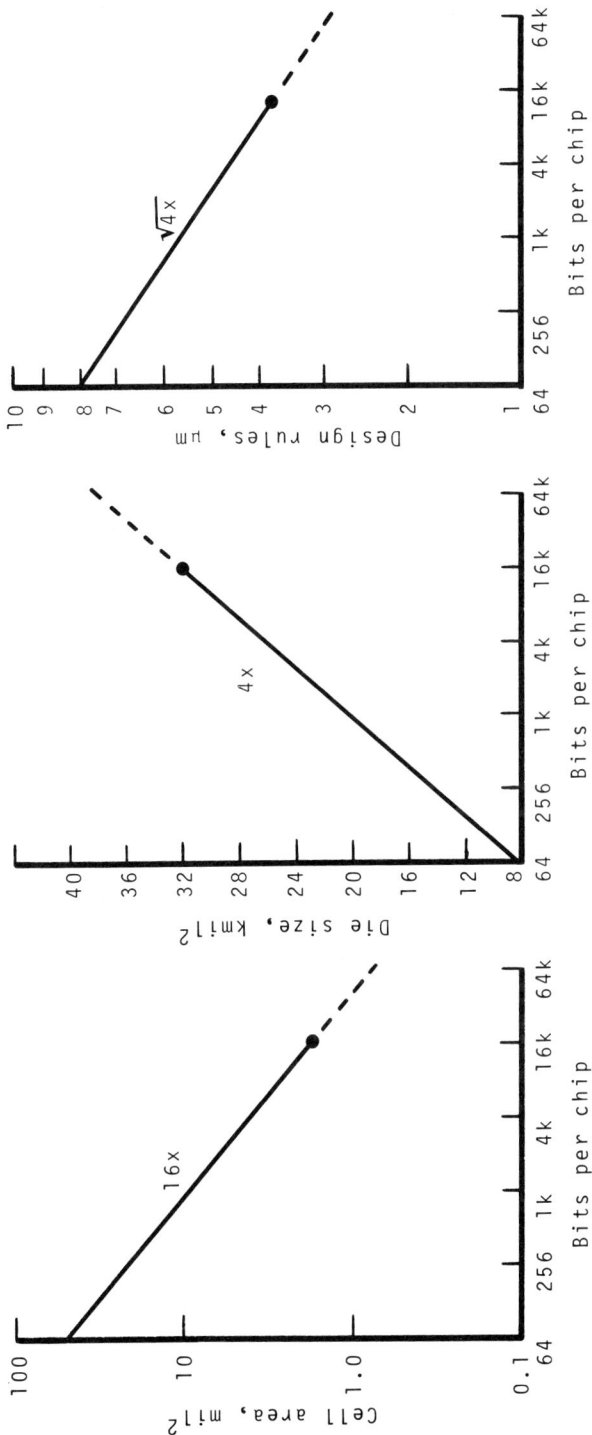

Fig. 2.2. Main factors contributing to the phenomenon of Moore's law.

Summing up, a very large number of smaller and larger steps in design and technology brought about this phenomenon of Moore's law.

Other technical improvements were also obtained. For instance, along with shrinking circuit size, the speed of memory operation has increased steadily. Reliability was improved by better control of technology. And various circuit innovations facilitated the application of these memories in, for example, computer systems. Nevertheless, even though technical complexity and density of circuits on a chip have increased, costs of production have not risen proportionately.

The Learning Curve

Over the years the price of MOS-dynamic RAM circuits (in cents per bit) has been falling dramatically. This is an example of the so-called learning-curve rule, which states that the cost of a mass-manufactured product decreases by a constant factor every time the cumulative number of units produced is doubled. The Boston Consulting Group originally advised Texas Instruments to use the learning-curve theory in its product planning and pricing policies.(17) They suggested the use of "forward pricing", in which the price of a new IC is initially set lower than needed to recover design and manufacturing costs. The seller expects a resulting high usage of the IC and a large market share. The learning curve predicts a rapid decline in production cost, allowing a profitable operation later in the product's life cycle. Cunningham(4) gives some interesting examples such as Model T Fords, electric power generation, and steel production. He illustrates how MOS-dynamic RAM memory declined to 68 percent of the original price after each doubling of total volume (see fig. 2.3).

The main reasons for this impressive cost reduction are:

- The manufacturing method of the IC's is basically a batch-processing one. As discussed earlier, more cells have been packed per unit area of silicon (and hence per batch), yielding lower cost per cell.
- The silicon wafers on which circuits are made have been increased in size (diameters increasing from 2 in. to 4 in.), which, because of batch processing, decreases manufacturing costs per cell.
- The yield of the manufacturing process is increased; often beginning with less than 10 percent of the circuits passing final inspection, gradual improvements can increase this number considerably.

INTEGRATED CIRCUITS AND IMPACT

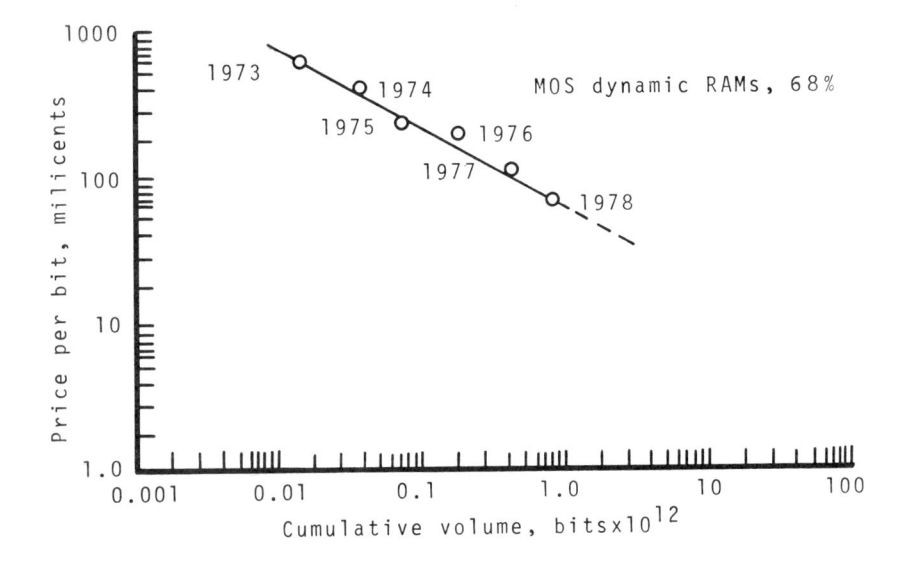

Fig. 2.3. Learning curve for nMOS dynamic RAM's.

- Improvements such as automation and large-capacity equipment have been developed and applied to large-scale manufacturing.
- Part of the expensive manual labor was transferred from the U.S. to low-wage (mainly East Asian) countries.
- Economies of scale begin to operate.

It should be mentioned that these cost reductions were counteracted by the necessary purchase of more sophisticated (and hence more expensive) equipment needed for the production of high-resolution circuits. In fact, capital investment needed in this type of industry is a major concern. The capital cost is so high and the economically useful life of the equipment is so short (because of the rapid improvements that come from new techniques) that a large investment is needed to obtain a turnover. (Capital investments typically amount to between 15 and 20 percent of turnover, Intel leading with 27 percent.) Nevertheless, as a net result the price per bit has come down sharply over the years, as figure 2.4 (from a lecture by Davis(5)) shows.

COST PER BIT
(CENTS)

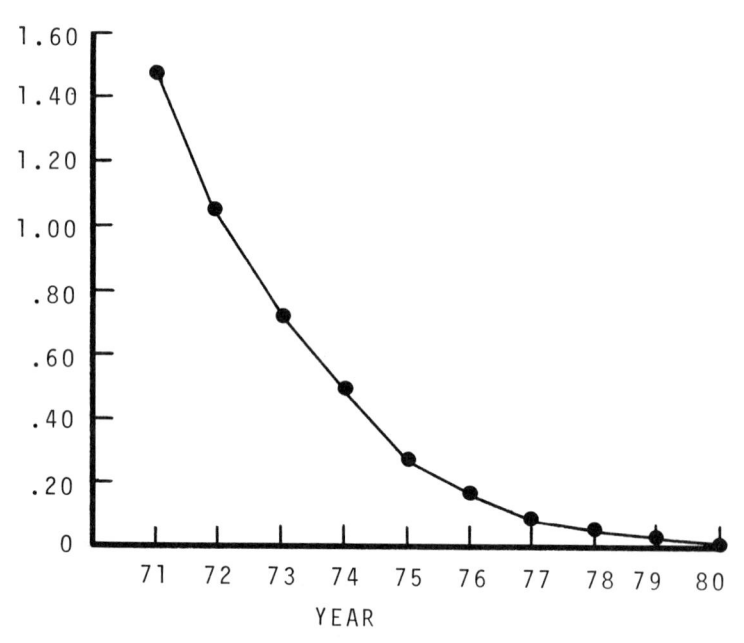

Fig. 2.4. Decrease of cost per bit for nMOS dynamic RAM.

The Generation Gaps

The foregoing discussion may have suggested that the transition from 64-bit to 16-kb RAM's has been a gradual and smooth process. However, this is not the case: the products come in generations, usually each one with a factor-four increase in capacity. This is illustrated in figure 2.5.(3)

This means that every two or three years a new generation of products must be marketed: the larger memory begins to be cheaper per bit than the foregoing generation. This gives companies which enter the market at an early stage a chance to set the standards, since they are usually not agreed upon prior to the introduction, and to capture a large market share. A large risk exists, however, because eventually only a few of these standards will be adopted by the industry.

INTEGRATED CIRCUITS AND IMPACT 15

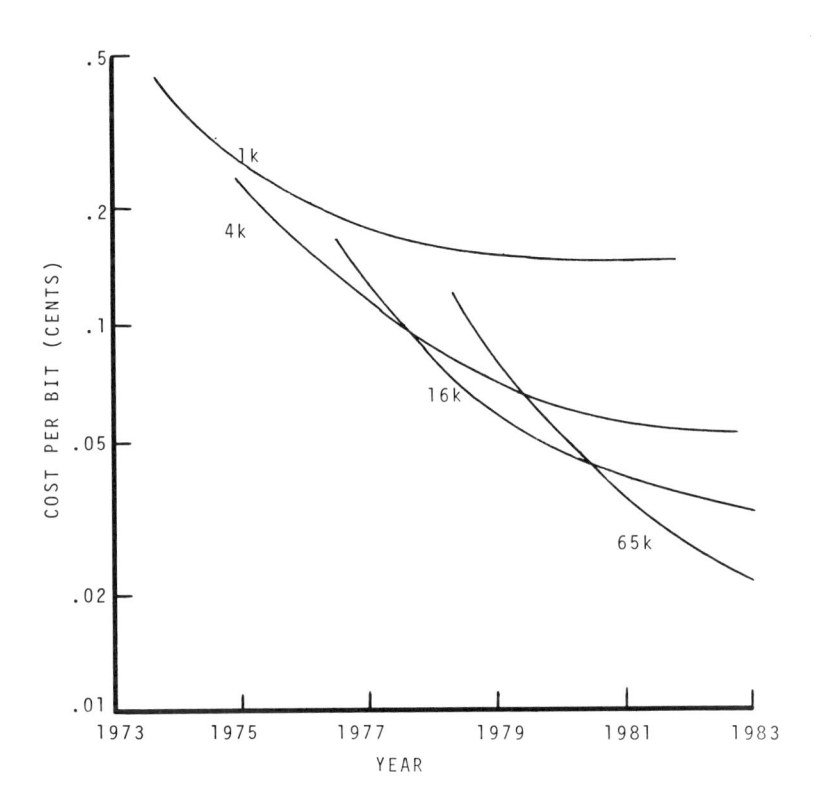

Fig. 2.5. Cost per bit for generations of nMOS dynamic RAM.

Summary

This example of the nMOS RAM illustrates major characteristics of the general-purpose integrated circuits as they are available on the market:

- A strongly increasing functional complexity.
- A steadily decreasing cost per function.
- A rapid succession of new generations of a certain functional circuit.

These phenomena are probably unique in the history of technology. This is certainly true regarding the speed with which this process occurs. Also unique is the immense impact which these technological changes have on other branches of electronics. In this respect we shall now investigate three important applications: computers, telecommunications, and control.

COMPUTERS

The development of computers(6) has been closely linked with that of integrated circuits in view of the predominant role IC's play in the computer's central processor and main memory. Hence the central processor's characteristics have been following the trends in IC's, resulting for instance in a much faster computing operation. In the main memory, magnetic-core systems have been replaced by semiconductor RAM's, yielding greater speed at lower costs. But significant progress has been made in other parts of the computer system. For example, the storage capacity of magnetic-disc memories has been greatly increased in the last decade.

The resulting decrease of the price/performance ratio and simultaneously increasing reliability have brought about extensive changes in computer usage. Electronic data processing, originally confined to big central departments, has become affordable for many divisions in large organizations and for many smaller organizations as well. Various types of smaller, more special-purpose machines have become available,(20) targeted toward the office, small business, real-time technical use, banks, and scientific desk-top calculation. Much additional hardware was developed for these applications, such as visual display units, magnetic cassette memories, small disc systems (floppy discs), and fast printers (e.g., daisy-wheel printers). Such smaller stand-alone systems can, where necessary, be connected to a large central computer. Data can be transferred to and from this central computer, whereas it can also carry out the more complex tasks or calculations.

One may expect this trend toward distributed-logic systems to continue. On the one hand there is a need in organizations for many special types of terminal or small computers. Various technologies may support such use, as, for instance, voice in-and-output and improved displays. On the other hand, an increased availability of telecommunications systems for data transfer will allow an improved communication pattern between different systems. In general, optimized architectures for such distributed systems still present design problems (as, for example, developing a system of distributed data bases).

This type of system with geographically distributed computers which are connected to each other should be distinguished from concurrent systems.(7) In these, several computers operate on the same task, which is divided between them. This can take the form of allocating the task over several interconnected computers (e.g., a computer working on a mathematical task can delegate the calculation of a necessary function to another computer). Alternatively, an array of (small) computers, connected only to their neighbors, can be used to attack a problem in a parallel manner. Both types of systems are still largely in the research phase, and indeed it is not clear how generally useful such designs would be. The associated programming techniques required are another area of study.

Large mainframes remain necessary, either to provide access to large data bases or to operate on complex problems. Special hardware-software combinations will be designed for the first task, the management of large data bases, and will include pooled software libraries. A different approach is needed for various scientific and technical problems (e.g., weather forecasting), where machines are required which are much more sophisticated than today's equipment.(8) Several approaches exist to increase computer power. The first one seeks to add a special fast parallel processor to the existing mainframe, which facilitates carrying out large numbers of vector multiplications.(21) In certain problems in which many of the necessary calculations can be programmed as vector multiplications this is a very useful approach, but for others it is not. It is also possible to build special high-speed computers for a specific task, such as data-base manipulation. A more generally useful approach is to increase the central-processor speed, for example, by choosing the fastest possible bipolar technology. Alternatives are the introduction of gallium-arsenide integrated circuits and Josephson junction logic. The last, under development at IBM for instance, particularly brings great promise for significantly increased performance of computers.(9) However, some technological problems remain to be solved. Apart from that, commercial aspects will undoubtedly play an important role in IBM's decision to apply Josephson junctions in commercial mainframes.

At the other end of the spectrum, small "personal" computers have emerged. Originally sold mainly to hobbyists, these inexpensive yet powerful systems are finding their way to small businesses for purposes of bookkeeping, text processing, inventory control, and so forth. In the future they will be introduced in the home, as prices come down further (in particular of peripherals like printers and floppy discs) and more "user-friendly" handling becomes available. It seems likely that such genuine "home computer" systems will become widely available in a few years.

Let us now turn briefly to computer software. The operating systems, developed by the vendor, have been changing drastically over the years. In addition to the conventional batch processing, the computer has come to accept time-sharing processing of a number of terminals. More recently, operating systems have been constructed which are optimized for interactive use of computers (e.g., the Bell Laboratories' UNIX system).

In addition to the operating system, the vendor usually supplies a number of utility programs, for instance editing and debugging programs and compilers for several languages.

The user requires application software to treat specific problems on the computer. This can be produced by the vendor, the user, or an independent software house. Often large standard software packages are chosen which are adapted to the details of the user's need. Nevertheless this application software is a major source of costs, and tools have been developed for the (semi-) automatic generation of such programs.

In a broader sense, software costs associated with a large computer system often exceed those of the hardware. This problem is getting more urgent with the decreasing price/performance ratios which result from the technological improvement of the IC's. Additionally, it appears that the costs associated with maintaining and updating software are often unexpectedly high.

There are several courses to keep these costs under control. They include:

- The adherence to systematic software development procedures, usually requiring a detailed overall design before the writing of any program, the generation of good documentation, etc.
- The use of computer tools in programming wherever possible.
- The use of high-level languages such as Fortran, Pascal, Cobol, and ADA, which permit a compact and clear way of writing the programs.
- The application of certain programming techniques which structure the program as a sequence of well-defined modules, disregarding possible short cuts in favor of clarity and testability.
- The improvement of the portability of programs, allowing them to run on different computer systems.

Though admittedly progress has been made along these lines, the total software costs projected for the 1980's are staggering. Moreover, it will be difficult to find the manpower necessary to carry out the systems design and programming needed. This may become a major factor limiting the application of computer systems.

Summarizing, the technological development which occurred in integrated circuits has strongly influenced computer hardware. Price/performance ratios decreased and entirely new types of computers could be designed, resulting in a dramatic change in application patterns. For the great majority of computer uses software has become the dominant cost factor, requiring a new emphasis in economic considerations.

As a footnote to this discussion, IC technology that resulted in higher-performance and cheaper computers has made possible a less exigent use of these machines. Many programs need fewer programming skills than in the past, and in addition a variety of aids and tools have become available that are economically viable. Thus one might expect that in the future only a few very high-level specialists will be required to deal with new concepts, tools and so forth. The others who work with software would be experts in applications rather than in programming.

Novel, speculative research programs aimed at a merger of artificial intelligence concepts in computers (for instance, the Japanese "Fifth Generation" plan)(18), may lead to quite different programming concepts. However, there is still a long way to go before they can be adopted in commercial systems.

TELECOMMUNICATIONS

Telecommunications has changed drastically, though probably not always visibly to the general public, owing to the developments in integrated circuits. Fortunately, the change has coincided with the introduction of digital techniques in the telecommunication networks. In the past, networks like the telephone system or mobile radio for land and sea operated in an analog fashion. New insights about digital transmission, combined with new possibilities of digital IC's and later with fiber-optic transmission techniques,(10) have led to a revolution in telecommunications technology. Let us examine some examples.

Telecommunications implies the transmission of a signal over long distances. Hence, a continuous problem is guarding the signals against distortions in amplitude or in phase as well as the influence of noise and external disturbances. It is important to note that a digital representation of the analog signal can in principle be exactly reproduced after transmission. In contrast, the external effects on analog signals can only be minimized, for instance by proper feedback techniques.

Fortunately, it is possible to represent precisely an analog signal in a digital way.(11) For instance, a telephone signal with a 4,000-Hz bandwidth can be accurately represented by a series of 8-bit binary numbers, which are transmitted at a rate of 8,000 times per second. In this case the resulting bit rate will be 8,000 x 8 = 64,000 bits per second.

This particular digital representation of analog signals can be transmitted via a variety of means, such as a coaxial cable or a microwave radio link. During transmission, distortions and disturbances affect the signal, usually necessitating a fair amount of signal processing to retrieve the original signal. The theory of digital signal processing(11) has advanced rapidly during recent decades. Its application has benefited enormously from the introduction of microelectronics, which allowed a cost-effective implementation of the theoretical concepts.

A final important development to discuss is the emergence of optical communication techniques. Thin glass fibers can transmit light over long distance with low attenuation. This technology (see, for instance, Giallorenzi(10) for an introduction) is highly compatible with digital microelectronics. It presents for the first time in principle a very-wide-band signal-transmission capability with a cheap medium. Though current applications in telecommunications are still mainly for large-volume-traffic trunk lines, fibers have great potential in "wired city" concepts,(12) where individual homes are connected to wide-band communication systems. A number of experiments in this direction are in progress or planned, notably in Japan and France.

The change in telecommunications by the introduction of digital techniques could be illustrated by many examples.(16) Digitalization of the public telephone system is a good one; how is it changing with the advent of digital techniques?

The public telephone network, in a very simplified description, can be thought to consist of (see fig. 2.6):

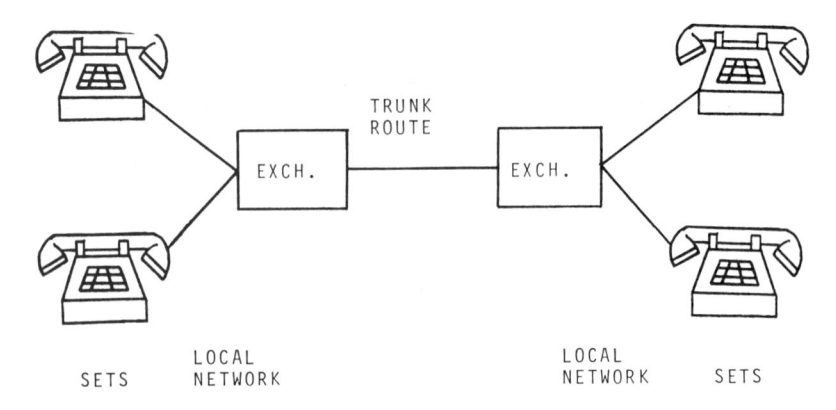

Fig. 2.6. Simplified diagram of the telephone system.

INTEGRATED CIRCUITS AND IMPACT

- The local network, i.e., the telephone sets at the subscriber's premises and the cable connection to the exchanges.
- A number of hierarchically organized exchanges, here for simplicity shown as only one.
- A trunk route carrying the high-volume traffic between the exchanges.

Digital techniques have been initially introduced in trunk lines. In the conventional "frequency-division multiplex" technique, a large number of telephone channels are modulated on a certain number of carriers with given frequencies. This technique allows the simultaneous transmission of many telephone conversations over one cable. In a digital transmission scheme, binary words belonging to different telephone channels are interleaved and serially sent through the cable, thus forming a high bit-rate stream. For instance, a bit rate of 44 Mb/s (44 million bits per second) in the trunk line allows the simultaneous (one-way) transmission of more than 672 conversations. At the receiving end, the samples belonging to each particular conversation are sorted out and combined to reproduce the original signal.

Obviously, this time-division multiplex mode of operation requires analog-to-digital conversion at the input stage of the trunk line, and the converse at the output. These converters are rather complex and expensive, but since they are relatively small in number (one set per trunk line), the cost is not particularly important.

In the telephone exchange, digital techniques were gradually introduced, beginning in control of the switching system. In the conventional electromechanical exchanges, each dialed digit sets a multiposition switch relay in the required position; the succession of dialed digits determines the path through the exchange. In the current generation, the dialed digits are first collected and fed into a central computer, which is programmed to set up the signal path through a number of electromechanical relays. The introduction of microprocessors makes it possible to decentralize a number of control functions. They are run on computer at a lower hierarchical level, the central computer getting a more supervisory and coordinating role. The advantages of this scheme are a greater modularity and simplicity of software.

So far, the exchanges still route analog signals via a large number of electromechanical switches ("space switching"). A major step forward is switching by using digital signals. In this case, an incoming digitally encoded telephone channel is time-multiplexed with a number of other channels. In a time-switching scheme, bits belonging to a particular incoming channel are selected from this bit stream and sent to the desired outgoing channel. Usually a combination of time and space switching is employed in modern digital exchanges. These digital techniques allow a vast reduction in hardware as electro-mechanical switches are replaced by a number of integrated circuits.

However, for this scheme to work it is necessary to transform signals on the incoming subscriber lines to a digital format, and vice versa. In the most direct scheme, each channel needs such a conversion circuit (Codec, coder-decoder) and a so-called subscriber-line interface circuit (SLIC). The latter allows the execution of a number of tasks like ringing or feeding power to the set to be carried out in the digital scheme.

The potentially large market for such Codec's and SLIC's has induced a number of merchant-IC and systems houses to develop such circuits, despite certain technological problems. They will be extensively applied in the next decade with increased deployment of fully digital exchanges.

It may be noted here that the concept of digital exchanges removes much of the complexity of the trunk-line end stations, because signals arrive in a neatly digitized and multiplexed format. This phenomenon of "integrated switching and transmission" requires a reorientation of the traditional strong separation of switching and transmission activities in the management structure of the administration and equipment manufacturers.

In addition, the local network may be digitized. This would simplify the interface problem with the digital exchange. However, the enormous investment in the local networks (cables, telephone sets), which is much larger than in exchanges and trunk lines, makes a quick replacement unattractive. A first step might be the use of a digital-transmission scheme between the sets and the exchange. A complication here is that the usual two-wire copper cables present in the local network are only marginally suited to transport the necessary digital signals. This is still an area of investigation, but it seems likely that solutions along this line will be introduced. A more attractive alternative would be to replace the existing local network with optical fibers, but it is unlikely that this will occur on any large scale in the foreseeable future.

With this description of the introduction of digital technology in the telephone system, the question arises, "What are its effects on the service provided to the general public?"

A major impact has undoubtedly been increased cost effectiveness of the telephone system for the user. Prices for better service have not increased with inflation.

Further, a number of smaller services can be provided, such as conference calling or call forwarding. More important, the use of telephone lines for connecting home terminals to central computers is being explored in different countries:

INTEGRATED CIRCUITS AND IMPACT

view-data systems like Prestel in the United Kingdom (many similar services exist in other countries under different names) connect the home TV to a computer, and the French project for an "electronic telephone book" provides an example of a specialized view-datalike terminal. It is certain that more uses for such terminals, for banking and electronic shopping for instance, will become available.

With this, controversies arise between the managers of the public telephone network, the potential supplier of terminals, and those interested in delivering services. In the French electronic telephone-book project, the French PTT (the government's organization for mail and telecommunication services) represents all three parties. In the U.S., the decoupling between the central network and the peripheral equipment is the subject of an antitrust lawsuit against ATT. On the other hand, the Federal Communications Commission will allow ATT to compete with other companies (like IBM) in the field of computerized-data communications.(19)

In addition to the public telephone network, specialized networks are becoming available for professional use. These data networks are often based on packet-switching principles (data are concentrated in a packet with the address of the destination and routed by means of a computer through a high-capacity network) rather than on conventional circuit switching (one connection is set up between the sender and receiver through which the data flow). Specialized services (including dedicated satellites) for business are being developed in the U.S. In Europe, the French PTT has similar plans. Sweden and West Germany have developed an improved telex service (Teletex). There are many other developments of data nets that illustrate the rapid progression of computer-communication schemes.

Thus, digital techniques, made possible by the economies of integrated circuits, are being introduced on a large scale in communications networks. This development has probably not attracted the attention of the general public as much as has the development of computer techniques. Nevertheless, it has provided a technical infrastructure for telecommunications which will greatly influence society. There clearly is a role for government to ensure proper use of these facilities.

CONTROL SYSTEMS

Control systems are used to regulate physical quantities: the flow of fluid through a valve, the speed of a motor driving a video recorder, the temperature in an oven. Such regulation entails the measurement of the physical quantity (pressure, rotation speed, temperature), followed by a comparison with some desired value and a subsequent action to minimize this difference (closing a valve, increasing the driving current of the motor, lowering the current through the heating spiral). Usually three units are needed: a sensor to measure the quantity, an electronic processor to determine the action to be taken, and an actuator to effect the change.

The theory of control systems has been thoroughly developed over the years. It is concerned primarily with stability criteria of the feedback loop, speed of operation, and sensitivity for disturbances. In the more recent past when digital computers became available, a digital version of the control theory was developed.(13) The ensuing control systems are in general more flexibly computerized than the analog systems, though several problems remain (e.g., quantization level, dead zone).

For instance, adaptive control systems can be designed in which the amount or nature of feedback can be varied according to need. Fault-tolerant control systems can be designed which survive the failure of one or more sensors.

Sensors utilize physical phenomena which transform a physical quantity into an electrical signal (e.g., a force can be transformed into a voltage by means of a piezo-electric material). A need has arisen for sensors that are compatible with integrated circuit characteristics and technologies.(14) Such sensors promise low-cost mass production and hence a broad application in microelectronics-based systems. It is sometimes advantageous to integrate signal-processing circuitry with the sensor on one chip, thus optimizing its performance. Special technologies will often be needed to achieve such combinations of sensor and IC.

Sensors with an output compatible with a digital circuit are desirable. For instance, effects that lead to a change in frequency (which can simply be counted) would be favored. However, many well-proven sensors with analog output exist, often combined with very sophisticated analog circuitry. The necessary analog-to-digital conversion can be performed separately.

It seems that the area of sensors has been somewhat neglected by most industries. It may prove to be a field with technical possibilities for new ventures based on an integrated-circuit type of technology. Such entrepreneurial companies may be in a position to serve the highly nonuniform market that exists today.

A digital computer is used in the control part of the control system. This was often a minicomputer suitable for real-time applications, but recently microprocessors have been used as a low-cost alternative.(15) Software design is usually based on extensive computer modeling and simulation of the system performance, using concepts of digital control theory discussed earlier.

A special type of circuit is often needed to drive the actuator when high power, high voltages, or high currents are needed. Several technologies permit the construction of high-power transistors, usually designated as DMOS, VMOS, ZMOS, and so forth.

The actuator itself can take on many forms, ranging from an electric stepper motor that opens and closes a valve to a piezoelectric ceramic rod which can make small movements necessary for tracking the replay head of a video recorder. Analog techniques are often required here. Also, in actuators and their driving electronics there is room for innovations and entrepreneurship.

Generally speaking, the cost reduction and performance increase of control systems made possible by the use of IC technology will make them applicable in many new areas. In consumer products, for instance, control systems are widely used in a broad range of applications from video recorders and video-disc players to automobile-engine systems.

REFERENCES

1. Abelson, P.H., and Hamond, A.L. (eds.), Electronics: The Continuing Revolution (Washington, D.C.: American Association for the Advancement of Science, 1977).

2. Hodges, D.A., "Microelectronic Memories," in Microelectronics, A Scientific American book (San Francisco: W. H. Freeman & Co., 1977).

3. Noyce, R.N., "Microelectronics," in Microelectronics, A Scientific American book (San Francisco: W. H. Freeman & Co., 1977).

4. Cunningham, J.A., "Using the Learning Curve as a Management Tool," IEEE Spectrum, June 1980, p. 45.

5. Davis, W.E., "High Technology Trade: A Prescription for Its Survival," Lecture Notes, S.I.A., 1981.

6. Mader, C., Information Systems, 2nd ed. (Boston: Charles River Associates, 1979).

7. Sugerman, R., "VLSI Computing: A Tough Nut to Crack," IEEE Spectrum, January 1980, p. 28.

8. Sugerman, R., "Superpower Computers," IEEE Spectrum, January 1980, p. 28.

9. Marcus, M.J., "Some Systems Implications of a Josephson Junction Technology," IEEE Proceedings 69 (1981), 404.

10. Giallorenzi, T.G., "Optical Communication Research and Technology," IEEE Proceedings 68 (1978), 744.

11. Rabiner, L.R., and Gold, B., Theory and Application of Digital Signal Processing (Englewood Cliffs, NJ: Prentice-Hall, 1975).

12. Carne, E.B., "The Wired Household," IEEE Spectrum, October 1979, p. 61.

13. Franklin, G.F., and Powell, J.D. Digital Control of Dynamic Systems (Reading, MA: Addison-Wesley, 1980).

14. Barth, P.W., "Silicon Sensors Meet Integrated Circuits," IEEE Spectrum, September 1981, p. 33.

15. Farber, G., "Status of Hardware and Software for Microcomputers," Conference on Digital Computer Applications to Process Control, Duesseldorf (New York: Pergamon, 1980).

16. "Survey Telecommunications," Economist, Aug. 22, 1981, p. 4.

17. Linvill, J.G., in: U.S.-Japanese Competition in the Semi-conductor Industry, Seminar Report, May 1981, Japan Society, New York, and Japan Society of California.

18. "Japan's Strategy for the 1980's: A Fifth Generation: Computers That Think," Business Week, Dec. 14, 1981, p. 94.

19. Uttal, B., "How to Deregulate AT&T," Fortune, Nov. 30, 1981, p. 70.

20. "Moving Away from Mainframes," Business Week, Feb. 15, 1982, p. 78.

21. Bernhard, R., "Giants in Small Packages," IEEE Spectrum February 1982, p. 39.

22. Schaefer, D.H., and Fischer, J.R., "Beyond the Supercomputer," IEEE Spectrum, March 1982, p. 32.

3
Microelectronics: Three Application Areas

The examples of microelectronics given in Chapter 2 may perhaps seem to have only a remote bearing on daily life. In fact, however, these technologies form the basis for many applications. Computer technology is relevant for many video games played at home. Telecommunications may change office life considerably, and control techniques may form the base of industrial robotics.

These three subjects - consumer electronics, office systems, and factory automation - may further demonstrate the pervasiveness of microelectronics in our everyday environment. But they show also the enormous industrial importance of microelectronics in our society. This is not only because the consumer sector provides a substantial market for microelectronics-based equipment. It will become evident from this chapter that a few key functions of our economic infrastructure, in particular the office and the factory, depend critically on the availability and quality of microelectronics technology. Thus, a U.S. microelectronics capability has become a key element in future growth potential, and as such it has become an area of public policy concern.

CONSUMER ELECTRONICS

In "conventional" electronic consumer equipment such as radio, TV, and Hi-Fi equipment, microelectronics has primarily replaced discrete devices like resistors, capacitors, and transistors with analog IC's. This has made the apparatus cheaper, more reliable, and lighter, often designed with better specifications. Also, several new concepts such as video tape recording, video discs, and portable color cameras have been made possible by integrated-circuit technology.

Digital IC's initially were applied in the control area of the equipment. For instance, a microprocessor was used in the interface between a video recorder's circuitry and the user control buttons. Additionally, new digital functions were added to existing apparatus, giving rise to new combinations such as the radio-alarm clock.

Later on, certain analog signal-processing circuitry was replaced by digital IC's, making possible, for instance, digital tuning in FM radio. Currently, there is a clear trend toward increased application of digital signal representations.(1,2) An example is the use of digital tape-recording techniques in the record industry. A logical extension of this is the proposed digital audio disc, which would replace the present (analog) records. The information on this disc is read by a solid-state laser optical system, and processed further by digital means, for instance to correct errors in the disc. Digital processing might also be applied to adapt the Hi-Fi system to the room characteristics, for example, by adding a certain amount of reverberation.

However, some caution should be exercised regarding the drive toward digitization in consumer electronics. For many applications, analog IC techniques provide simple, cheap solutions that are difficult to surpass with digital techniques. This is certainly true when further improvement of signal quality does not make sense. Also, most signal sources such as broadcasts of radio and TV are in an analog form. Though it would be quite possible to transform them into a digital format, this would require new standards and large investments by both senders and receivers, and so seems unlikely to happen soon. Rather, the transmitted signal may be improved or additional information may be added to it, while it is kept compatible with existing standards. Another approach to better quality could be upgrading the receiving equipment. For instance, the expected advent of a relatively cheap solid-state memory capable of storing one full TV picture may present the possibility of improving picture quality considerably.

Entertainment and information are beginning to reach the home in new ways. For instance, a TV signal may be broadcast with some additional digital information.(3) After decoding in the receiver, a number of pages of "teletext information" can be displayed on the screen, by use of a small command box. This service is now available under different names in a number of countries.

A TV set equipped with a similar decoder and command box, and an interface (modem) with the telephone network can be used to receive digital information stored in a central computer.(3) This system is also being offered on a trial

MICROELECTRONICS: APPLICATION AREAS

basis in several countries (e.g., Prestel in the United Kingdom).
Cable distribution systems are gaining widespread popularity. In addition to the normal broadcast TV they may deliver special channels devoted to particular subjects. Cable-TV systems can in principle also be used in a two-way mode. This gives novel possibilities for interactive video schemes, as well as for connection of the home to banks, shops, or other service and retail firms.
Finally, direct-broadcast satellites(4) are under construction by France and West Germany. These will beam TV programs and Hi-Fi radio (in Germany) directly to the individual's home (or to the cable system), which needs only a small outdoor dish antenna to receive the signals.
Some of the schemes described above raise serious questions of a nontechnical nature. For instance, who is authorized to determine the information allowed on a cable system? How can one cope with the "spillover" of one nation's satellite TV programs to neighboring nations? These and many other issues require a rethinking of some of the traditional patterns of information distribution.
A further example of the use of digital electronics is in several household appliances which were hitherto entirely electromechanical, e.g., clocks in automatic coffee machines, microprocessors in washing machines, microprocessor-controlled servosystems in sewing machines, and electronic control of heating equipment.
We can expect much more sophisticated use of microcircuits in the future.(34) New functions will be developed, instead of the simple replacement of existing functions currently in use. A washing machine might be programmed to continue washing until the laundry is clean rather than for a predetermined time. Energy management in the home can be optimized by electronic means. Home security, too, will benefit from a more intelligent combination of the signals from various sensors to cope with the current major problem of false burglar alarms.
Up to this point, we have discussed primarily the potential influence of microelectronics on existing home equipment or on new but related developments. However, some novel digital devices have already entered the home. The first one, the digital watch, has largely replaced mechanical devices as a low-cost, high-performance alternative. The second one, however, the pocket calculator, was essentially a new aid, since the existing electromechanical calculating machines and slide rules were used by limited groups, mainly in professions. It is interesting that the markets for both devices are largely in the hands of companies formerly not active in the consumer market.

In the wake of these new devices many other developments took place. Electronic game equipment, first in connection with the TV set, later in stand-alone versions, has been developed. The range of sophistication is wide, from simple children's toys to fairly complex chess computers. Educational aids have become available, as for arithmetic or spelling; likewise travelers' aids such as pocket translators.

A major new entrant is the home computer, though small-business owners and enthusiastic hobbyists were the initial users. It seems probable that this type of equipment will further penetrate the home, as it becomes more user-friendly and as certain applications develop such as access to central data banks.

New terminals for the telephone network will also be offered, either by extending current telephone functions (for example, shortened dialing and call forwarding) or by presenting entirely new possibilities (the French electronic telephone book, a terminal to "look up" telephone numbers stored in a central computer but which can also be used for more general view-data applications, or terminals for electronic banking and shopping).

Many other novel applications of microelectronics in the consumer sphere can be cited, either as isolated devices or fitting into large systems. In the latter case, electronic money is an intriguing possibility now being examined in France. In an advanced scheme the user would receive a monthly credit card in which the amount of money to which he is entitled is recorded in read-only memory. Purchases, recorded and added in a second memory, would be allowed until the issued value was reached.

To sum up, microelectronics has been introduced into consumer equipment at a revolutionary rate and there is every reason to assume that the process will continue. However, as the diversity of equipment in the home continues to grow, the need for some kind of order will become apparent. The user can benefit fully from the range of available functions only if some compatibility between equipment components is established, if the number of interfaces remains limited, and if duplication of functions is minimized.

However, it seems hardly necessary to require that a certain apparatus be able to communicate with every other apparatus in the home. It seems more logical to assume that a certain clustering into systems will occur, with only a rather loose coupling between the different systems.

Specifically, the equipment clustering may take the following form:

- A home electronics center will combine most quality video and audio functions.

MICROELECTRONICS: APPLICATION AREAS

- A home information center will develop around the home computer, including its terminal function, which will be able to handle information, to communicate, and to supply major educational services.
- A home control center will be mainly responsible for regulation of energy consumption, supervision of domestic appliances, and security.
- Portable equipment, partly stand-alone and some with an interface with one of the other systems, will cater to personal needs, in particular for outdoor entertainment and information.

It will certainly be a long time before such a clustering is reached. Inhibiting factors are numerous, including the incompatibility of various vendors' equipment, the lack of communication standards in general, and the nonexistence of an information-transmission medium in the house. Nevertheless, it seems likely that a scenario like that sketched above will develop over the course of time, but only after an initial phase of more growth and diversification.

Electronics and the Automobile

Until recently, the use of electronics in the automobile was limited to entertainment (car radio, cassette deck) and communication (mobile radio, citizen's-band radio, radiotelephone). But electronics is now being extensively applied to the engine control of American cars.(5, 27)

Originally, federal and state antipollution regulations spurred R&D efforts into the application of microelectronics for controlling engine operations. These efforts were enhanced after the oil crisis, when federal law began to require average levels of fuel economy in all cars sold in the U.S. In recent years the general public has turned toward more fuel-efficient cars, reinforcing market pressure on the automakers.

There seems to be no viable alternative for the U.S. automobile industry to the application of microelectronics if requirements of both pollution limitation and fuel economy are to be satisfied. Microprocessor-based electronic control systems developed so far use some measurements, such as oxygen content in the exhaust and temperature of the cooling liquid, to regulate the engine operation. The algorithms used depend on the properties of the particular car produced, such as the car body, transmission, and engine type. To cope with these variations, GM developed a control system with a programmable read-only memory, which is programmed with data which are relevant for a particular car.

Some major problems in the application of electronics in autos are the adverse operation conditions like vibration, temperature, and disturbance by electromagnetic fields. Moreover, electronics is a relatively new phenomenon in the carmaker's mechanical world, which has led to resistance against electronic solutions by designers and to the problem of training the service organization in an unknown technique.

Nevertheless, it seems likely that more electronics will be applied to the car. Present designs will be further advanced and the number of functions will be increased. This development will probably be spurred by an increased commercial interest in electronic features that can be offered.

In the near future three different electronic systems for cars will be standard: an engine control system; computer-type functions, such as calculations of actual fuel consumption or miles to go at current gas level; and entertainment units, perhaps combined with the former system.

Additionally, antiskid brakes, adaptable suspension, and collision-avoidance devices are being introduced in selected expensive cars. They may reach the mass market in a more distant future.

The architecture of the systems will probably be changing considerably from the current centralized "box under the dashboard" design. A distributed system may be attractive, certainly when other types of communication structures, for instance so-called buslines, become feasible.

THE OFFICE

The office has been identified as a primary host for the introduction of electronic equipment.(6) The prime reason for this trend is the small gain in productivity of an office worker relative to that of a factory worker in the last few decades. Additionally, the ratio of office workers to blue-collar workers has increased owing to the changing nature of the manufacturing process and growth of the service sector. Seeing that investment in capital equipment per office worker is very low, the question arises as to whether electronics can provide the means to increase productivity in this area.

Office work can be categorized into various job levels, for instance:

- Clerks (users of files, information storage and retrieval).
- Secretaries (typing, agenda organization, contact with managers).
- Middle management (daily supervision, budgeting, planning).

MICROELECTRONICS: APPLICATION AREAS 33

- Top management (long-term planning, representation, overall decisions).
- Professionals (general data use and report writing in different fields: scientific, engineering, financial, medical, and law).

Clearly, every category has its own type of work for which particular electronic aids might be useful. We shall initially provide a rough sketch of some of the currently available equipment(7) and services and their introduction into an organization.

In the last decade, equipment for text generation has become popular. Dictating equipment has spread quickly, though actually the equipment seems to be used infrequently. Electronic typewriters have been equipped with some additional features, such as a line memory. More recently, word processors have been developed into a variety of forms, from a stand-alone unit to a part of a minicomputer-controlled system. They may be equipped with a full-page cathode-ray tube or with a more limited display; with a magnetic card memory or with a floppy disc. The software may contain only simple correction features or sophisticated report-editing facilities.

Microfiches and microfilm are receiving increased attention for filing and retrieval of documents in view of their high-density storage characteristics. The combination of a computerized file with some key data regarding the documents and a computer-addressable microfiche file seems to be a particularly attractive option.

Communication methods both inside and outside the office include several private services (internal mail, office telephone with private exchange (PABX)), and public services (mail, courier, telephone, telex, data networks). Some new services, such as data transmission and facsimile transmission, have come into operation.

Computer systems have been used extensively in the office for routine tasks such as data handling, invoice handling, or financial administration. Initially, this was done only in large centralized computer centers of major companies. When less expensive specialized business computers became available, smaller firms and departments of large companies became involved. The emergence of the home computer brings such administrative aids to very small businesses and private persons.

Information systems (usually terminals connected to a data bank) have found many applications. Some operate entirely within one organization (inventory control, management information); others utilize external sources (scientific literature references, legal and financial data).

Generally speaking, the goal when introducing electronic equipment is to increase office productivity, for instance in terms of text produced per worker or in time needed to retrieve a document. Another, more important goal is enhancement of the effectiveness of various categories of personnel (e.g., better client service provided by clerks, or improved professional judgment by easy access to critical data).

Currently, various equipment and services for the office are rather disjointed. One might expect, however, that a process of integration(8) will occur, presumably in a number of steps:

- Equipment and services are offered to assist the office worker; for example, new aids are introduced in the same pattern of office activities.
- This leads to a proliferation of equipment that calls for coordination. Equipment can be chosen that is matched to the tasks performed by the office to obtain an efficient flow of work. Here, organizational changes will be initiated: the office is going to do things differently.
- A next step will be the integration of the various equipment into a novel office environment. For instance, text generation, copying, transmission, filing, and retrieval could all be done electronically in one system. The same system, with some additions, might be used also as a management information system. A combination of the two tasks in one system obviously leads to changes in the procedures adopted by an office. Often new organizational forms will be required.
- This can then lead to the performance of new tasks which were hitherto not feasible. This, again, will lead to changes in the organization and may well generate novel requirements for equipment.

It is unclear how long these processes will continue before stabilization occurs. Tensions may arise between a highly organized, and therefore probably centralized, office environment and the flexibility needed to adapt quickly to changing external conditions. One may expect, though, that the potential of increased speed of communication and availability of detailed information will be attractive to an organization. Its decision-making process will be speeded up and the reaction time shortened.

The actual development of technology suggests that such a course is within reach, certainly from the point of view of hardware. An exhaustive discussion of the coming technology is beyond the scope of this book. However, let us review some of the technological knots that have to be untied, even if the task of the system is clearly defined.

MICROELECTRONICS: APPLICATION AREAS 35

- A major issue is the architecture of future integrated office systems. Numerous questions arise such as how to exercise control over the system, where to put intelligence, and how to distribute storage.
- The structure of the internal electronic communication systems within an organization is a major issue. Networks will certainly carry voice and computer data, possibly video (picturephone, conference TV) or very high bit-rate high-resolution picture data. Bit rates differ substantially per application: approximately 64 kb/s for voice, 1-2 Mb/s for video, 5-10 Mb/s for high-resolution pictures. Various bit-rate reduction techniques exist which may find application here.(32) The design of the network can take a variety of forms, usually on-site private networks are connected via the public network.
- Interestingly, the proposed architectures for on-site networks reflect the backgrounds of the various companies involved. For instance, the telephone companies propose an improved (digital) telephone network controlled by a private telephone exchange. Mainframe vendors envisage a large central computer to which a number of intelligent terminals are attached. Minicomputer makers often propose some form of busline over which various stations can communicate. At the moment no compatibility exists, though efforts are being made to arrive at a standardization.(9)
- More possibilities are becoming available in the public networks.(10) In the U.S., specialized data nets for business are developing, including systems which rely on direct communication between sites via satellite. They will allow a combination of services, e.g., telephone, electronic mail, data transmission, and video conferencing. The European development, dominated by the various PTT's, is much slower. International compatibility of data nets is a major problem, but it is in the hands of large organizations with vast experience in such matters. It will come, although slowly.
- Functions are being combined into one piece of office equipment. Electronic reading of typed text, combined with character recognition on one hand and high-quality printing linked with a copier on the other, presents the necessary interfaces between the worlds of paper and electronics. If the document is not conducive to the use of such techniques, high-speed facsimile can be used. Many combinations of equipment can be designed, but only a few will evolve as practical and cost-effective.
- Terminals for specific user groups are becoming available.(25) Ideally, the equipment should be adapted to the needs of the user groups. For instance, for clerical

staff working long hours with a display, optimal picture quality is required. For managerial terminals, very simple handling is desirable.(11) The development of properly adapted equipment will require a considerable amount of research and development. Many of the desirable features, method of interaction for instance, can only be found after extensive experimentation in a realistic environment.
- Various technologies are being developed which may be attractive for use in the office. Optical memories for instance, provide a very dense storage medium.(12) The application of voice input or output is a special feature. Voice-command operation and speech recognition are feasible techniques. However, a system that really "understands" normal spoken language and transforms it into typed text is in the distant future. It would require, in addition to understanding words and sentences, a considerable amount of linguistic analysis to determine the context of the words in order to arrive at the proper spelling.
- Major problems arise over the authorization of access to, or manipulation of, certain data. In many countries laws have been passed guarding the use of private data banks. Also, transborder traffic of data may be restricted.
- Security of data transmissions is another issue.(35) Various cryptographic systems have been proposed. There is concern about the security of the Data Encryption Standard (DES), accepted by the U.S. government.(13) Also, new intriguing concepts such as "public key" systems(14) are arousing much controversy.
- Computer-fraud prevention poses several legislative and technical problems. This subject will receive increasing attention in the next decade.

We have now stressed that the current technology provides quite a number of possibilities for the introduction of productivity-improving equipment in the office. The cost of such equipment has been reduced to make it economically attractive. Microelectronics has played a major role in achieving this: the capabilities of the equipment have been greatly improved, the size has been reduced, and the price has declined.

Nevertheless, from a technical point of view there is ample room for further improvement. In particular, the input and output functions will probably undergo drastic changes in the next decade. More generally, software may prove to be a critical factor. This is true in regard to the necessary new concepts such as novel types of interactive software, but the

availability of sufficiently skilled analysts and programmers may become a bottleneck.
However, rather than the availability of technology per se, the main issue will be how it can help to satisfy an organization's need for information, and how it can help make the necessary decisions in an increasingly complex environment. This need can be fully understood only when the complete range of relationships among people, organization, and technology is considered. Consequently, any introduction of technology into the office requires a carefully designed strategy of change.(15) Little experience is available. It seems likely that a slow process of learning has only just begun: where it will lead in the next decade is unpredictable.

FACTORY AUTOMATION

In discussing the influence of microelectronics in the factory, it is useful to distinguish between process industry and manufacturing industry. The former is highly capital-intensive and was an early adopter of computer control; the latter is much less homogeneous in its use of capital goods.
The process industry is already highly automated. Control theory has been widely applied in extensive computerized control systems. These systems usually consist of several hierarchically ordered layers.(16) Micro- and minicomputers are present at lower levels to physically measure and control the process and interact with the operator. Larger computer systems are applied in supervisory plant control systems and management information systems.(17)
In the manufacturing industry, automation has proceeded more slowly. Of course, many products such as light bulbs or TV picture tubes are produced with large specialized machinery. A major part of the manufacturing industry, however, involves batch processing or assembly activities which require a substantial amount of human labor.(18)
A strong drive exists to replace this human labor by automated machinery. The economics of such a move has become increasingly attractive: labor costs have been rising, whereas the costs of computer-controlled machinery have been declining. International competition leading to increased productivity is a further stimulus. Moreover, the advantage of consistent quality achievable in an automated production process is now receiving increased attention.
A major drawback of extensive automation has been the rigidity that results from the usual mechanization: a line intended to manufacture a certain product cannot easily be switched over to make a different product. Thanks to

microelectronics, this is going to change. Computerized numerically controlled machines, often with a number of tools incorporated into one unit, have been developed and can be programmed for a variety of jobs. Robots,(19,20) are being introduced which can move objects, handle tools, or assemble parts, all under program control. Automated material transport and warehouses have become available. (A more extensive discussion of these "flexible automation" and "robotics" concepts is not presented here. It may suffice to note that these techniques are often considered a determining factor in an industry's ability to produce cost-effectively at a high quality level.)(33)

Initially such equipment was introduced on a stand-alone basis, for instance a single numerically controlled lathe to perform a certain part of the production process. In the course of the time the components are being connected into groups of machines performing certain more complex tasks. A layer of communication between the units and the groups is needed to coordinate the various tasks that comprise the full production flow. A coherent system of machinery, internal transportation, assembly, testing, packing, and forwarding can then be developed.

This "bottom-up" approach is complemented by a "top-down" trend. The factory computer, initially used mainly for several administrative jobs, is increasingly used as a tool for order control, inventory control, and planning. The two automation trends meet in an intermediate logistics layer, which translates planning and other production data into an actual flow of work through the factory.

Computer-aided design (CAD) methods have initially gained widespread use in electronics design. Similar methods are being applied to the mechanical domain.(23) CAD methods must be blended into the automated factory process, since they generate the design data about the products to be produced. Here, the problem of a general data base arises: data describing the product will be used for many different purposes, such as ordering, manufacturing and testing. A multiusable data base has to be designed, requiring careful coordination with various procedures in use throughout the organization.

Evidently, the introduction of such far-reaching fabrication technologies is a slow-moving process.(21) International competition is a major driving force. Japan in particular has taken the lead with government-sponsored programs aiming at the development of a coherent technology base in this field.

Future developments will include robots which can feel and see. Tactile sensing can be made already at acceptable costs. Seeing and subsequent pattern recognition (e.g., of mechanical parts scattered on a conveyor belt) is still a slow and rather costly process. One may expect that special

MICROELECTRONICS: APPLICATION AREAS

integrated circuits for picture processing and pattern recognition will become available. A drastic cost reduction may ensue, leading to a strong increase in the application of sophisticated robots.
Additionally, improved techniques for automatic measuring and testing will be needed. For rapid, high-rate data transmission in the factory, optical fibers are attractive in view of the "noisy" environment. Easily adaptable programs are needed to instruct robots and numerically controlled machinery. Specialized high-level languages (and perhaps in the future artificial intelligence systems) will find applications here.
Summarizing, microelectronics has made a considerable contribution to enabling factory automation to become feasible and attractive. Systems integration, however, will remain a major problem for some time. Also here, extensive software programs will be needed at various levels. The linking of such programs will be a considerable task. Further, progress in integrated circuitry for picture processing will lead to a new wide range of applications of robots.

ASSESSMENT OF THE U.S. POSITION

A complete balanced and detailed judgment of American capabilities in the field of microelectronics is beyond the scope of this presentation. (See also an Office of Technology Assessment study.)(22) Some comments on the competitive position of the U.S. in today's world in the areas we have just described may be useful for later discussions.
In telecommunications, the American industry has a leading position in technology. In particular, the fields of satellite communications and data networks are far advanced in the U.S. home market. The Department of Defense has been a major factor in the development of these techniques. However, outside the U.S. several Japanese(31) and European companies hold important market shares. A major factor here is the influence of many national governments, via their PTT organizations, on the industrial R&D efforts and on the division of the market.
The American position is also fairly strong in control technology. The process-control equipment market is dominated by American companies, with some European competition. The emerging CAD-CAM market is also strongly pushed by U.S. companies.(23) In "flexible automation" and "robotics" for assembly-type manufacturing, however, Japan seems to have taken the lead with well-concerted programs in relevant areas of fabrication technology.(26,28,33) The corresponding American efforts seem to be less coherent.

This is disturbing for the U.S., in view of the importance of manufacturing techniques for the productivity and hence the competitive position of a wide range of industries.

Currently the U.S. dominates the world computer market. European efforts to establish a computer industry have been unsuccessful. Japan is under way with a broad program aimed at creating an indigenous computer industry. It encompasses both hardware and software development, whereas very advanced concepts are also getting much attention.(24,30) Confrontation between the American and Japanese products is about to begin seriously. The Japanese hardware may prove to be attractive, but the American computer companies may hold a distinct advantage in software.

In the field of office automation, American companies have also taken the lead. Assisted by a relatively flexible public policy regarding telecommunications, the American computer and telecommunication industries are focusing on this market. European companies may have a fair chance in their own markets, which usually develop several years after the American market. The force of the Japanese industry is undetermined, though several technological strengths (printers, copiers, displays) deserve attention.(24)

Finally, the American position in the more mature consumer electronics industry has been weakened during the last decade by Japanese and other Far East competition.(22,29) It has made major inroads in the American market and put the local industry on the defensive. The European industry, though also under stress from Japanese competition, is still an important factor.

In summary, the overall world position of the American microelectronics industry is very strong. True, in several areas severe competition now exists or is developing, notably from Japan, but in general this can be viewed as a healthy development: American industry will be forced to innovate and increase its productivity in order to avoid an experience similar to that of the American consumer electronics industry.

It is evident that the American microelectronics industry strongly depends on the American position in integrated circuit technology. A healthy, first-rate merchant industry is essential, as is a high technical level of the captive activities. The leverage of the IC technology position is appreciable: though IC's usually account for about 5 to 10 percent of the equipment value, they determine its technical performance and total price. Many factors contribute to a nation's position in a certain technology. Clearly the technical ones predominate, but others can influence the health, direction, and scope of the enterprise. We shall examine some of these key considerations as we follow this discussion.

MICROELECTRONICS: APPLICATION AREAS

REFERENCES

1. Bernhard, R., "Higher Fi by Digits," IEEE Spectrum, December 1979, p. 28.

2. Society of Motion Picture and Television Engineers, "Television Technology in the 80's," S.M.P.T.E., 1981.

3. Jackson, R.N., "Home Communications I: Teletext and Viewdata," IEEE Spectrum, March 1980, p. 26.

4. Harrop, P., et al., "Satellite Communications II: Television for Everyone," IEEE Spectrum, March 1980, p. 54. Also: Pritchard, W.L., and Kase, C.A., "Getting Set for Direct Broadcast Satellites," Ibid., August 1981, p. 22.

5. Rivara, J.G., "Microcomputers Hit the Road," IEEE Spectrum, November 1980, p. 44.

6. Mokhoff, N., "Office Automation, A Challenge," IEEE Spectrum October 1979, p. 66.

7. Manual, T., "Automating Offices from Top to Bottom," Electronics, March 10, 1980, p. 157.

8. Zisman, M.D., "Office Automation: Revolution or Evolution?," Sloan Management Review, June 1978, p. 1.

9. Mokhoff, N., "Local Data Nets: Untying the Office Knot," IEEE Spectrum April 1981, p. 57.

10. Kaplan, G., "Three Systems Defined," IEEE Spectrum October 1979, p. 42.

11. Business Week, "Will the Boss Go Electronics, Too?," May 11, 1981, p. 106.

12. Bulthuis, K., et al., "Ten Billion Bits on a Disk," IEEE Spectrum August 1979, p. 26.

13. Sugerman, R., "On Foiling Computer Crime," IEEE Spectrum July 1979, p. 31.

14. Diffie, W., and Hellman, M.E., "Privacy and Authentication: An Introduction to Cryptography," IEEE Proceedings 67 (1979), p. 397.

15. Grunsteidl, W., "Man and Automation in the Office," 14th International TNO Conference, March 19, 1981, Rotterdam.

16. Evans, L.B., "Industrial Uses of Microprocessors," Science, March 18, 1977.

17. Sugerman, R., "Electrotechnology to the Rescue," IEEE Spectrum, October 1978, p. 53.

18. Olling, G., "Experts Look Ahead to the Day of the Full-Blown Computer-Integrated Automatic Factory," IEEE Spectrum, October 1978, p. 60.

19. Zermeno, R., et al., "The Robots Are Coming, Slowly," in T. Forrester (ed.), The Microelectronics Revolution (Oxford: Basil Blackwell, 1980), p. 184.

20. Munson, G.E., "Robots Quietly Take Their Places, Etc.," IEEE Spectrum, October 1978, p. 66.

21. Bessant, J., et al., "Microelectronics in Manufacturing Industry: The Rate of Diffusion," in T. Forrester (ed.) The Microelectronics Revolution (Oxford: Basil Blackwell, 1980), p. 198.

22. Office of Technology Assessment, U.S. Industrial Competitiveness, A Comparison of Steel, Electronics, and Automobiles, Washington, D.C., 1981.

23. Bylinski, G., "A New Industrial Revolution Is on the Way," Fortune, Oct. 5, 1981, p. 106.

24. "Japanese Strategy for the 80's: Information Processing," Business Week, Dec. 14, 1981, pp. 65,74,78.

25. Uttal, B., "Xerox Xooms Towards the Office of the Future," Fortune, May 18, 1981, p. 44.

26. Niff, R., "Japanese Rush to Robot Production," Electronics, Oct. 6, 1981, p. 87.

27. Holusha, J., "Detroit Plugs into Electronics," New York Times, Dec. 9, 1981.

28. Lerner, E.J., "Computer-Aided Manufacturing," IEEE Spectrum, November 1981, p. 34.

29. "The U.S. Consumer Electronics Industry and Foreign Competition," Northwestern University, PB 80-184807 (1980).

30. "Japan's Strategy for the 80's: A Fifth Generation: Computers That Can Think," Business Week, Dec. 14, 1981, p. 94.

31. "Japan's Strategy for the 80's: The Coming Assault in Communications Markets," Business Week, Dec. 14, 1981, p. 98.

32. Haskil, B.H., and Steele, R., "Audio and Video Bit-Rate Reduction," Proceedings IEEE 69 (1981), p. 252.

33. "Japan's Strategy for the 80's: The Push for Robotics Gains Momentum," Business Week, Dec. 14, 1981, p. 108.

34. Lerner, E.J., "Micros In White Goods," IEEE Spectrum, April 1982, p. 50.

35. Bernhard, R., "Breaching System Security," IEEE Spectrum, June 1982, p. 24.

4
Integrated Circuits, Technology

Any further analysis of microelectronics as a source of technical change requires a basic understanding of its driving force, integrated circuits. In this part of the book we set out to describe the basic manufacturing technologies, the major product classes, and the industry structure. Only after this can we attempt to discuss with greater understanding aspects like the strength of the national technology base or international competition.

Integrated circuits are available for an incredible variety of functions, produced in many specific models in several different technologies by a wide range of suppliers. Referring to figure 4.1, integrated circuits can either be classified in technologies, mainly bipolar and MOS, or in functions, such as memories and amplifiers. The function of a certain product and the technology in which it is executed are usually interrelated. For instance, a very high-speed microprocessor for military applications may preferentially be built in some bipolar technology. Conversely, an IC for an inexpensive watch may utilize a low-dissipation MOS technology.

In the broad spectrum of merchant producers, many have specialized in a certain type of IC or have built up strength in particular technologies. Captive producers also develop specific strengths, usually around circuits which are critical to their systems. Several companies, including some new ventures, specialize in services like custom design of IC's.

Any analysis of the role of integrated circuits as "building blocks" of microelectronics requires some detailed knowledge of technologies, products, and industry and their interrelations which will be developed in this part of our discussion.

INTEGRATED CIRCUITS, TECHNOLOGY 45

Integrated Circuits: Technologies

Bipolar
- TTL and Schottky TTL
- Emitter Coupled Logic
- Integrated Injection Logic
- Integrated Schottky Logic

MOS
- pMOS
- nMOS
- CMOS
- DMOS

Special
- GaAs
- SOS
- MESFET

Other
- Bubble memories
- Josephson Junction devices
- Surface Acoustic wave devices

Digital
- Memories
- Microprocessors
- General Logic
- Specific Logic
- Gate Arrays

Analog
- General purpose
- Special purpose
- Interface

Other

Integrated Circuits: Function

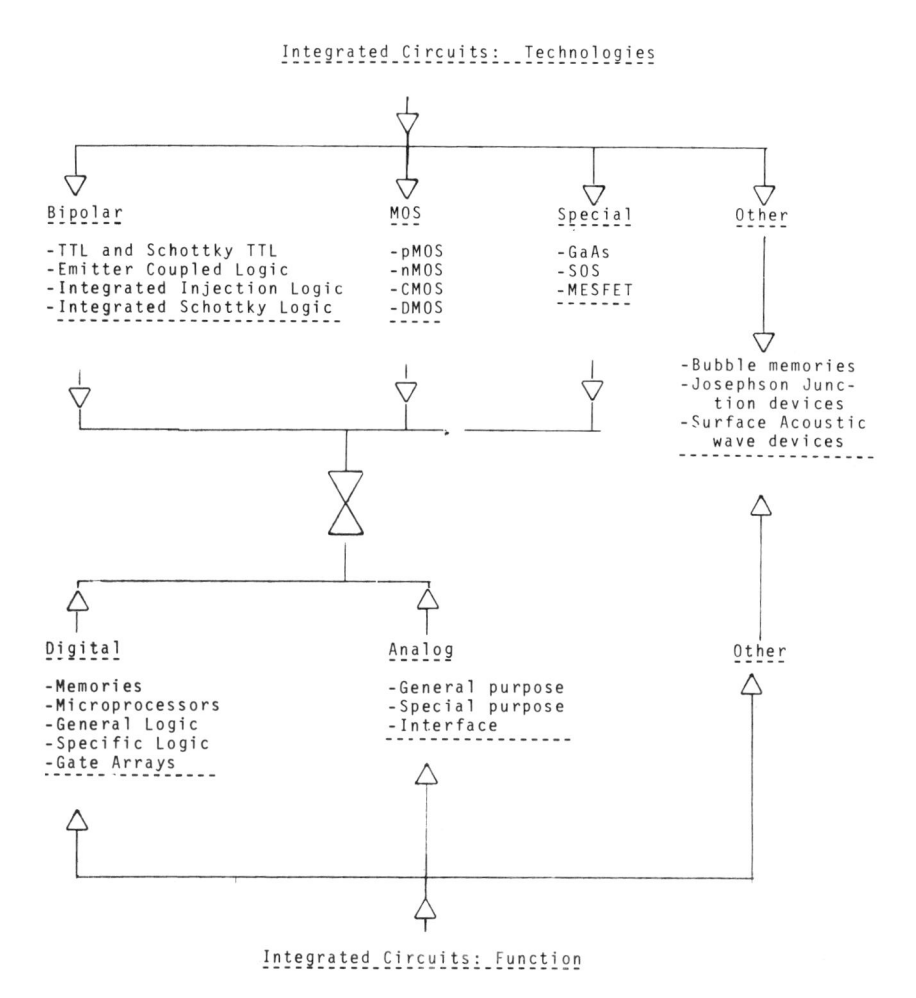

Fig. 4.1. Integrated circuits: technology and products overview.

In this chapter, we discuss several important questions, such as: what is an IC and by what means and methods is it designed and manufactured? Subsequent chapters will deal with products and with the industry.

IC TECHNOLOGY

Several main technologies are in use in the integrated-circuit industry. They are applied in making integrated circuits, consisting of many thousands of tiny transistors on a small (typically ¼" square) piece of silicon (the chip). These manufacturing technologies require highly precise operations, based on optical and mechanical techniques as well as on the procedures of solid-state physics. Many hundreds of identical IC's are made simultaneously on the surface of one thin circular wafer of silicon.

In this chapter we sketch the physics of transistors and integrated circuits and describe the basic technology in use for the production of IC's. After focusing on the most common bipolar and nMOS technologies, we briefly review the current status of the technology.

TRANSISTORS AND INTEGRATED CIRCUITS

Transistors

The transistor was invented at Bell Laboratories in 1948 by John Bardeen, Walter H. Brattain, and William B. Shockley. In a simple form it consists of three layers of monocrystalline semiconducting material such as silicon. Each layer has been treated differently: by substituting other atoms with a different number of valence electrons for some of the silicon atoms in the crystal lattice selected, one can dramatically change the electrical properties of the silicon crystal.

If atoms like phosphorus are substituted for silicon, a process called doping, an excess of electrons is created in the material (n-silicon). Conversely, substitutions with atoms like boron create an electron shortage (p-silicon). In transistors the three layers are respectively of p, n, p material or n, p, n material. The properties of the three layers can be chosen in such a way that an electrical current passing through the transistor can be strongly influenced by a much smaller current (or voltage) applied to the middle layer. This makes the transistor an "active component," in contrast with "passive components" like resistors, capacitors, and the like: a small voltage or current can be used to change a large current (amplification).

INTEGRATED CIRCUITS, TECHNOLOGY

Figure 4.2 illustrates this description. An introductory paper by Meindl(1) explains the operation of transistors more completely. Many textbooks are available on semiconductor physics; a typical treatment is the work by Gibbons.(2)

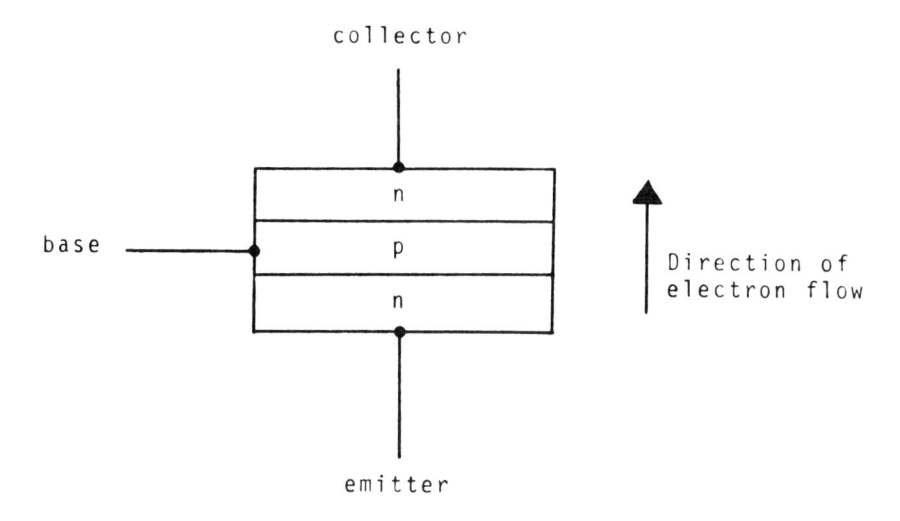

Fig. 4.2. Bipolar npn layer transistor.

In the mid-1950's a "planar" transistor was developed: the p, n, p regions were made on the surface of a silicon crystal (figure 4.3). The various regions could be obtained by selectively doping parts of the surface by masking those areas that must remain unchanged.

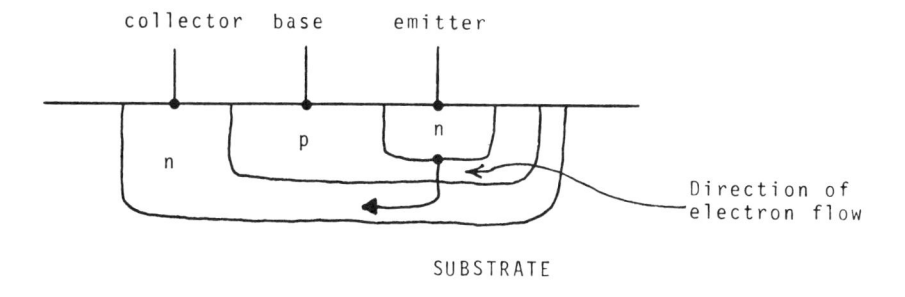

Fig. 4.3. Planar npn transistor.

Integrated Circuits

Later on, techniques were developed to make a number of transistors on the same surface, each electrically isolated from the other. Also, an insulating layer with suitable holes was applied on top of the transistors, allowing metal paths to be made to connect the various transistors (figure 4.4). This led to the integrated circuit, first realized in 1959 by J.S. Kilby of Texas Instruments and R.N. Noyce of Fairchild.(3)

Though the first integrated circuits consisted only of a few transistors, refined design and production techniques quickly increased the number of transistors per circuit. The following classification is often used; small-scale integration (SSI, less than 100 components per IC); medium-scale integration (MSI, 100-1000 components per IC); large-scale integration (LSI, 1000-100,000 components per IC); and very large-scale integration (VLSI, more than 100,000 components per IC). The first VLSI circuits have now been introduced on the market.

Bipolar and MOS Transistors

So far, we have used as an example an integrated circuit of the "bipolar" type: the basic transistor consists of separate areas of n and p type materials. It is the oldest type of integrated circuit and is still very important. However, another device, the Metal Oxide Semiconductor transistor (or MOS transistor) has found a wider application.

As indicated in figure 4.5, electrical current can flow between two regions of n material, called the source and the drain. This current can be influenced by the voltage on an isolated electrode or gate situated directly above the current path. A negative voltage will cut it off; a positive voltage will let it pass. (MOS transistors can also be made with p-channels. Instead of the "normally on" transistor discussed here, "normally off" types can be made too.)

A comparison of bipolar and MOS transistors leads to the following observations:

- The MOS transistor can be made smaller, allowing for a higher density of integration.
- The bipolar transistor is faster.
- MOS manufacturing technology is simpler and thus less costly.
- The bipolar transistor is better suited for analog operation.

In the marketplace, sales of MOS-IC's have been growing faster than that of bipolar IC's. This trend has been helped

INTEGRATED CIRCUITS, TECHNOLOGY

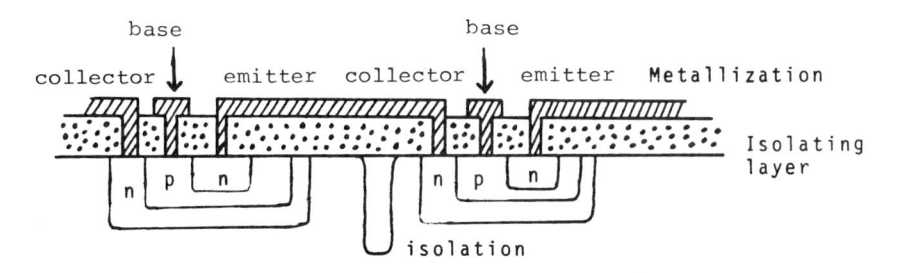

Fig. 4.4a. Bipolar integrated-circuit principle.

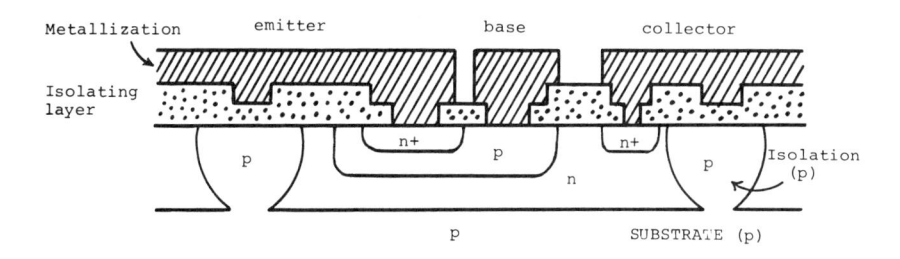

Fig. 4.4b. Bipolar integrated circuit: actual cross section.

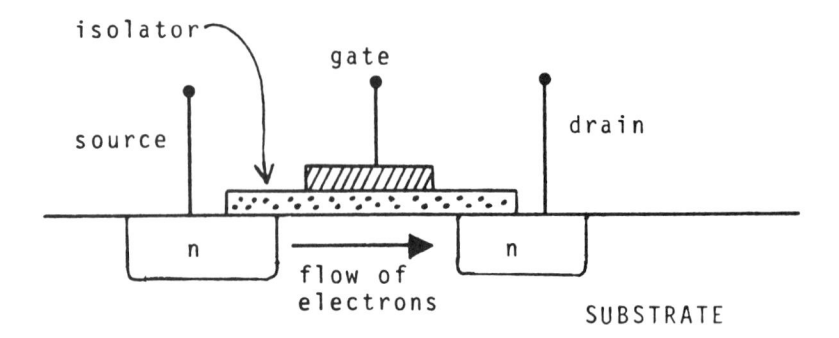

Fig. 4.5. nMOS transistor.

by the fact that the speed disadvantage of MOS-IC's diminishes with decreasing transistor size, as is the case in the current trend toward VLSI. Though the market lead of bipolar IC's has been taken over by MOS-based circuits, the former may be expected to keep a strong position in several important areas.

Classes of IC's

Usually, the broad classes of bipolar and MOS-IC's are further subdivided. In bipolar IC's several different topologies of transistors and diodes are used, each having special advantages regarding speed or integration size. For instance, Transistor-Transistor Logic (and the newer form which includes so-called Schottky diodes) has become so popular that an extensive "logic family" of compatible circuits has been built for general-purpose applications. Very high speed can be obtained with emitter-coupled logic (ECL), however at a relatively low density. Much higher densities can be obtained with Integrated Injection Logic (I^2L) and Integrated Schottky Logic, albeit at a somewhat reduced speed.

In MOS, different technologies can be distinguished, such as nMOS and pMOS already mentioned. The former has become the major vehicle for high-density circuits such as random-access memories. In recent years Complementary MOS (CMOS) gained importance. By properly combining n and p MOS transistors in one circuit, the steady-state dissipation of a circuit can be greatly reduced. This is important in many applications such as battery-powered equipment. In addition, special technologies such as Double-Diffused MOS (DMOS)

INTEGRATED CIRCUITS, TECHNOLOGY   51

must be mentioned. Circuits made in these technologies can withstand high dissipated power or unusually high operating voltages.
  Many books, such as Colclaser's, deal in greater detail with the subject of integrated-circuit classes and their properties.(4)

BASIC MANUFACTURING TECHNOLOGY

Highly complicated technology is involved in the manufacturing process for integrated circuits,(2,4) but we will focus only on the basic steps involved.(5)
  Very pure crystalline silicon can be produced in the form of a rod with a diameter of typically 3 in. to 5 in. Thin round slices of silicon are sawed from this rod. After polishing and cleaning, these wafers are ready for use in the IC production. Each wafer will accommodate several hundreds of identical IC's.
  As discussed earlier, several areas of each individual IC must be doped with materials like boron and phosphorus. Isolating layers have to be applied on top of them. Conductors, connecting various doped areas, have to be applied. All this is done step by step, with all IC's on the same wafer being treated simultaneously. The process is schematically illustrated in figure 4.6.
  First, however, in some cases (especially in bipolar technology) a thin, highly regular layer of silicon is grown on the wafer. Then the silicon wafer is oxidized at a high temperature, resulting in a thin layer of insulating oxide covering the surface. This in turn is covered with a thin layer of photosensitive material, usually called resist.
  A glass plate, called the mask, on which the pattern to be transferred has been etched, contains the hundreds of identical patterns corresponding to the individual IC's. It is placed on top of the wafer and irradiated with a light source. The resist that has been exposed is then developed and removed. Then the oxide can be removed from the places not covered with resist by means of an etching process. Finally, all remaining resist is removed.
  A wafer is now obtained, covered with oxide, with (in each separate IC) several openings that give access to the silicon surface. This wafer is now placed in a furnace in, for instance, a boron atmosphere. At an elevated temperature, boron diffuses into the silicon crystal lattice, giving rise to doped regions. In this process, the temperature, dopant concentrations, and process duration have to be carefully controlled to achieve the desired concentration profile of dopant. When this stage is finished, the remaining oxide can

52   U.S. MICROELECTRONICS INDUSTRY

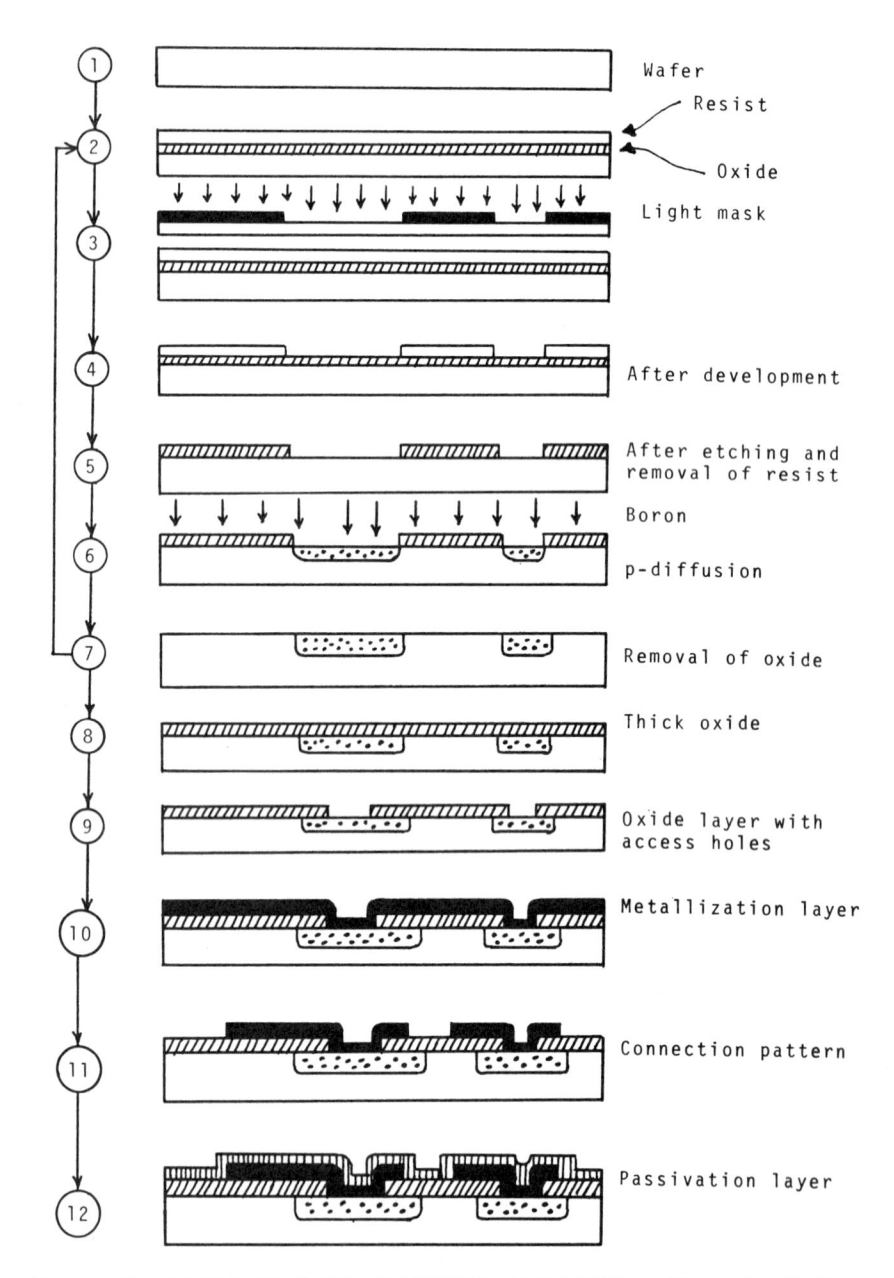

1. Wafer
2. Resist / Oxide
3. Light mask
4. After development
5. After etching and removal of resist
6. Boron
7. p-diffusion
   Removal of oxide
8. Thick oxide
9. Oxide layer with access holes
10. Metallization layer
11. Connection pattern
12. Passivation layer

Fig. 4.6. BASIC MANUFACTURING PROCESS. For simplicity only one single IC is shown on the wafer. Actually many hundreds are produced simultaneously, as explained in the text. See also fig. 4.7.

be removed. In this way the p-regions on the wafer are made, the latter being ready for further treatment.

One can now continue to make the n-regions by repeating the process: start with oxidation and resist application, use a second mask to make the n-regions accessible, and diffuse with phosphorus. Usually a number of such repetitions is necessary to complete all necessary diffusions.

After these steps have been completed, a thick layer of oxide is evaporated over the entire surface. By means of the techniques just described, holes are made in this oxide, opening up parts of the silicon surface. A thin metal layer is then evaporated over the surface, making contacts with the silicon through the holes in the thick oxide. The desired pattern of conductors is made, again using the lithographic techniques in etching away the undesired metal. Finally, a protecting (passivation) layer may be applied over the whole circuit.

Remember that the above description is highly simplified: an actual manufacturing process may only faintly resemble the one discussed here. However, the major steps in every process will be easily recognizable, though they may actually be performed differently. For instance, the fabrication of MOS-IC's requires the construction of a gate. This may be done with similar techniques, but other possibilities exist.

IC DESIGN AND MANUFACTURING

IC Design

The process of designing and manufacturing a new integrated circuit begins with drawing up a specification: what is the task the IC or set of IC's is supposed to perform? In the architectural phase several alternative solutions are compared, divisions in functional blocs are considered, preliminary estimates of area needed for transistors and interconnections are made, input and output pins are assigned.

When the major architectural decisions have been made, the actual circuitry is designed. Usually it involves a number of representations with a gradually decreasing level of abstraction. This could, for instance, be as follows: block diagrams and logic equations, logic diagrams, stick diagrams, and finally the actual IC layout.

Several computer tools are usually applied, for instance, to check designs against the logic equations, for modeling of transistors and circuits, for running checks on the timing in the circuit, and for simulation of circuits. Computer-aided design systems, available from several companies, are applied in the actual mask-design phase, where transistors and interconnections are laid out on the available surface.

Mask Making

When this design phase is ended, magnetic tapes are generated which contain information for the production of masks (one for each pattern step discussed earlier). They are used to make a set of "reticles," glass plates on which the patterns are etched at a magnification of about ten times the size of the actual IC. These reticles are used to make the actual masks on glass plates on which the patterns have the same size as the IC itself, at maximum several tens of mm^2. The pattern is repeated in a regular cycle to fill the full surface of a wafer (which has a diameter of 5 to 10 cm), allowing the simultaneous production of many copies of the same IC on one wafer. This procedure, requiring several specialized optical machines, is schematically shown in figure 4.7. It is usually carried out in a specific mask-making center.

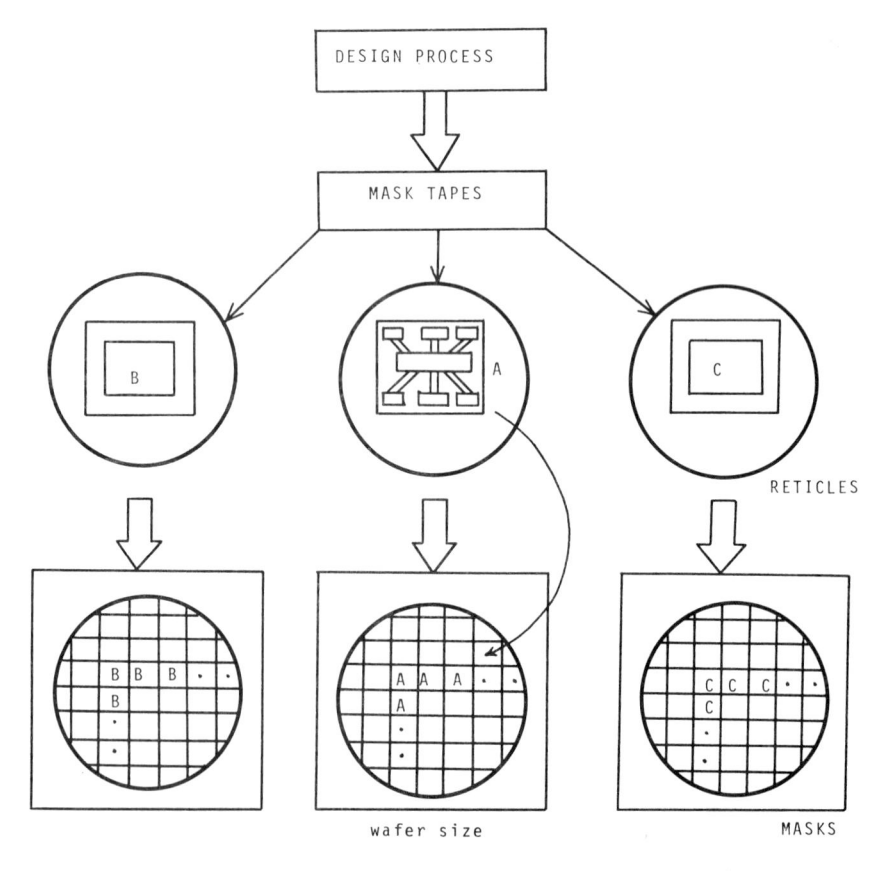

Fig. 4.7. Mask production.

INTEGRATED CIRCUITS, TECHNOLOGY

Wafer Fabrication

Now comes the actual manufacturing process of the IC, along the lines described earlier. Actually, copies of the first set of (master) masks are used for obvious reasons. They are either placed directly on the wafer (contact printing) or at a slight distance. Special projectors are applied in which the alignment of the successive patterns is a major source of concern. Usually a number of wafers with the same circuit are processed simultaneously (batch processing). The process conditions in the various steps must be kept under close control; the intermediate steps, such as cleaning and manipulating the wafers, must likewise be performed with great care. This part of the manufacturing process still depends largely on manual operations, though computer control of processes and computerized "track record" systems are increasingly being introduced. These sets of operations are concentrated in the "wafer fabs" of the IC house, in which much of the technological know-how of a particular company is concentrated and protected. Much capital is invested here, in equipment as well in ultraclean rooms.

Testing and Mounting

Finally, the wafer-processing phase leads to a wafer with a large number of copies of the same circuit. They undergo a functional test on the wafer, usually with very complex testing equipment. The wafer is then cut into "chips" or "dice," each containing one circuit. They are glued on a metal frame which contains the pins of the IC package. The circuit is connected to the pins with very thin gold wires (wire bonding). A house of plastic or ceramic is provided around the IC and the pins are bent into correct position. The finished IC may be tested again and is finally ready for shipment.

This part of the manufacturing operation requires a good deal of manual labor. Many IC houses have established facilities in cheap-labor countries, where the whole mounting process is concentrated. The Japanese industry has opted instead for automation. A similar course was followed by several captive producers in the U.S. For instance, IBM has highly automated its production lines.

Finally, a measure of the quality of the manufacturing process is the yield of good circuits. It often starts with a few percent at the beginning of a production. Obviously, every effort is made to push the yield to obtain a linear decrease in the costs per chip. Yield figures are among the best-kept trade secrets, since they directly reveal the profitability of a certain operation.

U.S. MICROELECTRONICS INDUSTRY

CURRENT STATUS

Let us review the present capabilities of the technology and some newer manufacturing techniques which are superseding those contained in the rather classical description above.

Design Techniques

IC design techniques have not kept pace with the possibilities provided by technology. Many computer-aided design (CAD) techniques have been developed to model, simulate, and lay out small- and medium-scale IC's.

However, applying these programs to LSI or VLSI circuits, which have so many more components, requires unwieldly amounts of computer time. Thus, new concepts in CAD for these complex circuits are needed. Much attention is currently focused on this subject, but progress is rather slow.

Wafer Preparation

Better command of the production technology of crystalline silicon has resulted in an increased diameter of the wafer. Whereas some years ago a 2-in. wafer was standard, currently built waferfabs use 4-in. wafers. There is an obvious economic advantage: the same set of process steps yields four times as many chips. Additionally, a better quality of crystalline silicon allows larger chips to be made, since the areas that are free of faults increase.

Lithography

Lithographic techniques have been greatly improved. Currently, line widths of around 3 micron can be routinely produced in factory processes. This involves partly a continuous upgrading of the projectors for contact printing and proximity printing. A newer optical method is the wafer stepper, where a single large mask is projected on the wafer. Mechanically moving this image stepwise over the wafer surface gives the necessary repetition of the circuit pattern, the so-called step and repeat method.

Optical methods are limited to line widths of the order of the wavelength of the light used ("normal" light about 0.5 micron, but deep ultraviolet can be used with special lens systems). Electron-beam pattern generators have been developed in analogy with electron microscopes, which surpass

INTEGRATED CIRCUITS, TECHNOLOGY

optical microscopes in resolution. They are potentially capable of a much higher resolution, but their operating speed is limited. Hence their use is generally restricted to the mask-making process. Direct writing on the wafer is technically possible, but is not yet a cost-effective way of making high-resolution IC's.

Process Technology

Process technology has been enriched with techniques that offer a better process control. For instance, diffusion techniques used for doping (a time-consuming high-temperature process) can often be replaced by ion implantation. Here a beam of high energy ions is used to bombard the surface to be doped, resulting in an improved control of dopant concentration and profile. Moreover, doping can even be effected through a thin layer of oxide on top of the wafer. Wet chemical etching may be substituted by dry plasma-etching techniques. The latter offer again better control, especially since anisotropic etching can be obtained.

In the development of the manufacturing processes new materials are introduced. For instance, polysilicon instead of metals has been increasingly used for gates and interconnections. The relevant film-deposition techniques (evaporation, sputtering, growing of layers with chemical vapor deposition, etc.) have been continuously improved. All this has proved to be essential to reach the densities of present-day IC's in mass-production conditions.

Testing

Testing of the circuits is getting more complicated not only because of their reduced size, but also because their functional complexity has greatly increased. The way the circuit should be tested ought to be a major consideration in the design phase. Actually, many testing programs are still being developed at the time the circuit design has been finished. This is an unfortunate situation, which may improve when better links between computer-aided design, manufacturing, and testing will have been established.

Packaging

Most IC manufacturers do not have very specific packaging technologies. Usually circuits are mounted in standard plastic packages, when required by the user in hermetically sealed ceramic houses. However, the number of connecting pins increases sharply with growing circuit complexity. New techniques for packaging are needed in the VLSI era.(6) Also, IC's have increasingly to fit into assemblies other than the usual printed-circuit boards. There are various thin- and thick-film techniques, in which components are mounted on ceramic substrates. IBM has developed a multilayer ceramic substrate with a special connection technology for the IC. One may expect that such technologies will increasingly receive more attention. More compact equipment design, faster electronics, more power dissipation, and lower costs may result from further work along these lines.

REFERENCES

1. Meindl, J.D., "Microelectronic Circuit Elements," in Microelectronics, A Scientific American Book (San Francisco: W.H. Freeman & Co., 1977).

2. Gibbons, J.F., Semiconductor Electronics (New York: McGraw-Hill, 1966).

3. Noyce, R.N., "Microelectronics," in Microelectronics, A Scientific American Book (San Francisco: W.H. Freeman & Co., 1977).

4. Colclaser, R.A., Microelectronics (New York: Wiley, 1980).

5. Oldham, W.G., "The Fabrication of Microelectronic Circuits," in Microelectronics, A Scientific American Book (San Francisco: W.H. Freeman & Co., 1977).

6. Lyman, J., "Packaging V.L.S.I.," Electronics, Dec. 29, 1981, p. 66.

5
Integrated Circuits, Products

We have considered the design and manufacturing aspects of integrated circuits. What are the various products of the industry? As discussed in figure 4.1, integrated circuits can be classified in two major functional classes, analog and digital. In analog circuits, the value of the voltage or the current presented at the input varies continuously with time. It usually represents some physical phenomenon, for instance speech on the telephone or an image on TV. In digital circuits, only a series of discrete voltage (or current) values ("ones and zeros") are used. They can represent either numbers to be manipulated, as in computers, or digitized representations of analog signals, as in digital telephone transmissions. Generally speaking, analog IC's require a higher degree of mastery of circuit functions. For example, their amplification and phase response characteristics have to be accurately controlled. In digital circuitry, fairly wide ranges of voltage values will be defined either as a "one" or a "zero," making, for example, amplification requirements less severe. Speed and timing of the signals are of prime importance, however.

The 1979 U.S. production of integrated circuits is shown schematically in figure 5.1. The strong position of digital technologies, in particular MOS, is evident.

DIGITAL MEMORIES

Semiconductor digital memories consist of an array of separate cells. In each cell one information bit (a 0 or a 1) can be written and stored for some time. At a later stage this information can be read and used in further processing.

60   U.S. MICROELECTRONICS INDUSTRY

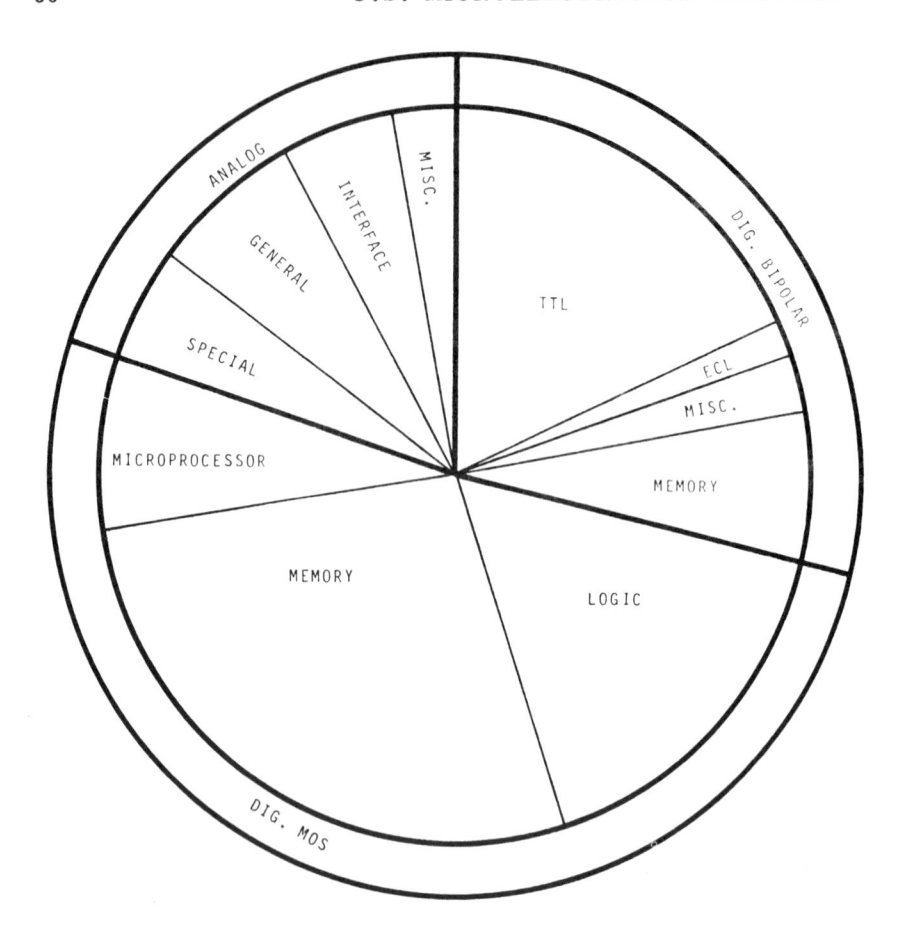

Fig. 5.1. I.C. production (sales value) in the U.S., 1979.

Source: Philips estimates.

Depending on the method of writing and/or reading, a number of different types of memory have been developed. An introductory discussion has been given by Hodges.(1)

Read-only Memories (ROM)

As the name implies, the content of this type of memory is fixed, and it is possible to perform only a reading operation. For example, a ROM may hold a fixed program to be used in a pocket calculator for a certain algorithm such as computing a square root.

INTEGRATED CIRCUITS, PRODUCTS

A ROM is often organized as a matrix of a large number of separately addressable words (for instance each 8 bits wide). When the address of a certain word is presented at the input circuit of the ROM, it gives as output the stored bit pattern of that word. The memory cells are rather simple: they contain a transistor connecting the input (word) line and output (bit) line if a 1 is contained, and no connecting device for a zero. Further, the chip contains the necessary circuits for such functions as decoding and coding of the input and output signals, and timing.

The contents of the memory (i.e., the distribution of the transistors over the cells), determined by the manufacturer, cannot be changed by the designer who applies the circuit in the equipment.

Programmable Read-only Memories (PROM's)

Functionally, a PROM is equal to a ROM. However, it can be programmed by the user who builds the memory into the system.

Organized in a similar way as the ROM, with a number of words that can be separately addressed, each of the PROM cells is provided with a transistor in series with a fusible link. The user of the circuit, using special equipment, can destroy the links at specific cells by applying a large current through them. As a result, some memory cells will still have the transistor connection between input and output lines, while others will not. The final circuit operates just as if a ROM were used.

The advantage of a PROM over a ROM is, of course, that the user can introduce his own programs; the disadvantage is the intrinsically higher cost per bit of the memory. The choice between ROM and PROM is ultimately an economic one. For large series it is advantageous to design masks and develop a special ROM; for small series this procedure is much more expensive than using a PROM and the necessary associated programming equipment.

Erasable Programming Memories (EPROM, EEPROM)

Though the PROM gives the manufacturer a greatly enhanced flexibility, it is programmable only once. In many applications it would be desirable to store a program and be able to exchange it occasionally for another program. Devices which allow this are sometimes called "read-mostly memories."

The memory cell in this case consists of a small, extremely well-insulated capacitor, on which a stored charge will be maintained for many years. This capacitor also forms the gate electrode of a MOS transistor, which is cut off when a charge is present and conducts if no charge is present. Thus, as in the preceding case, each cell contains either an operating or a nonoperating transistor, determining one of the two logical states.

This capacitor can be charged by injecting electrons, applying high-voltage pulses to the transistor. The charge can be erased again either by irradiating the memory with ultraviolet light, optically erasable PROM, or the use of electrical pulses, electrically erasable PROM or EEPROM.

A number of technologies have been applied for these devices, notably a Metal-Nitride-Oxide-Semiconductor structure (MNOS) and nMOS variations. However, fully satisfactory EEPROMs have not yet been made. They sometimes suffer from undesired side effects such as a deterioration in performance after repeated reloading.

Static Random Access Memory (static RAM)

In a Random Access Memory every cell of an array can be addressed for "read" and "write" operations, all cells having approximately the same access time. Each cell consists in principle of a couple of transistors in a two-branch circuit, flip-flop. This circuit has the property that either one branch or the other is switched on, again representing one of the two logic states. The circuit can be switched from one state into the other by applying a suitable electric pulse ("write" operation), whereas it is relatively simple to detect the state of the cell ("read" operation). When the circuit is in a certain state, it will remain so until a "write" operation is performed, hence the distinction "static" RAM.

Again, bipolar and MOS technologies have been applied for this type of memory. The memory chip usually includes decoders and coders for the input and output signals and timing circuits.

Dynamic Random Access Memory (dynamic RAM)

Here, every memory cell, addressable at will, contains only one transistor and one small capacitor. The charge on the capacitor, applied via the transistor, determines the logic state of the cell.

In comparison with the static RAM, one complication arises: the charge on the capacitor leaks away in some milliseconds, requiring a continuous regeneration of the memory. Charges on the capacitors are read at regular intervals, restored to their initial values, and brought back again onto the capacitors. To this end the chip contains a special regenerator circuit, in addition to the usual decoders and coders.

Apart from this added complexity, a dynamic RAM has the additional disadvantage that part of the time it is not accessible for "read" or "write" operations because regeneration is in progress. On the other hand, however, the memory cell is simple and hence very small. Especially in nMOS technology, the integration of dynamic RAM's is further advanced than that of any other type of circuit, which has led to the lowest price per bit stored. This advantage and wide applicability ensure a large market, making the dynamic RAM an ideal product for the manufacturers to recoup their massive investments needed for new technologies. Dynamic RAM's have been made in many technologies like nMOS and various bipolar versions.

Shift Registers

It is also possible to construct a memory with a "pipeline" organization. For example, a number of bits are serially entered into a FIFO register (first in, first out), from which they emerge after a certain amount of time. By feeding these bits back to the entrance, an amount of information, representing for instance a digitized television picture, can be stored over a long time.

Many variations on this type of memory exist, such as circuits circulating a number of bits in parallel or designs converting a serial input into a parallel output.

Applications and Trends

Memories are extensively used in the computer industry, though a multitude of other applications exist. The choice for a certain application depends on many factors such as price, size, speed, power dissipation, and compatibility with the additional circuitry. The requirements differ strongly, too: in a computer, a "scratch pad" memory may be small but must be fast, whereas a backup memory must be large but may be slower.

What are the indications for the development of the memory technologies of the future? (See also Posa(10) for a review.)

- From our discussion, we see that today's most advanced IC's in terms of number of components per chip are nMOS dynamic RAM's. It is economically attractive to strive for high densities, and it is technically possible to achieve this thanks to the highly repetitive nature of the circuit.

- One may expect that the trend toward further integration will continue; announcements of several manufacturers and experimental results of the Japanese VLSI laboratory are clear indications. Increasingly, redundancy is built into the memory: additional cells can be used to replace faulty ones. It is highly possible that unexpected problems will be encountered, such as the previously discovered effect of electron showers induced by alpha particles (from the materials surrounding the chip) which influence the charge accumulated in the cell. However, there seems to be no reason to expect that fundamental problems will soon arise.
- The major advantage of bipolar circuitry compared with MOS, namely speed, is eroding since MOS is getting faster as its dimensions shrink. Since MOS is inherently cheaper, bipolar-memory manufacturers are applying special technologies (such as Emitter-Coupled Logic) and circuit designs aimed at the highest possible speeds. Nevertheless, it seems apparent that in the second half of the 1980's the competitive edge of bipolar circuitry will be lost.
- There is still much room for innovation in erasable PROM's. Though the memory type is attractive, current solutions have more or less serious draw-backs. New companies may find an attractive initial product in this sector, provided some good new ideas are available.
- The application of CMOS technology(2) in memories is getting increased attention. It allows high circuit density at low dissipated power (when the circuit is idle; but when many accesses are necessary, the power consumption increases to levels customary in other technologies). Consequently, CMOS is not only well suited for battery-operated memories but is also attractive in very-high-density circuits where dissipation on the chip is beginning to become a limiting factor.
- Combinations of the memory types discussed above are likely. For instance, Electrical Erasable PROM's and dynamic RAM's may be equipped with supporting integrated circuitry making them look like static RAM's.(3) Such "pseudostatic RAM's" can be made compatible with microprocessors.
- In a number of applications, nonvolatile memories which do not lose their contents when the power is switched off are desirable. In such cases, magnetic bubble memories may be used, but in principle the Erasable PROM variety also possesses nonvolatility. Thus, improvements in this type of memory may lead to new applications where data have to be stored over long periods but need to be changed infrequently.

- It is uncertain whether multilevel logic, not operating in a binary way but for instance with 0,1,2,3 as "digits," will find applications. Intel has a 4-level PROM, in which three different transistor types are used.
- Other types of memory organization, such as associative memories, may gain importance. However, this will be determined more by the interest of the computer companies in using such new concepts than by the ability of the IC industry to develop them.

MICROPROCESSORS

The first microprocessor was marketed in 1971 by Intel. In one decade a tremendous growth in both market size and circuit performance has occurred. A review of hardware and software characteristics of microprocessors reveals the highlights of this development.

Microprocessor IC's

A microprocessor is a small "computer on a chip," a fixed "hardware" configuration which can perform a variety of tasks, specified by suitable "software" programs. Many different designs are available. Toong(4) gives an introductory discussion and more detailed treatments are also available.(5)

Microprocessor circuits have at least the following functions (usually on one chip):

- An appropriately coded program (stored in a separate memory) can be read and decoded in a series of subsequent instructions.
- Data (also stored in a separate memory) can be read.
- The operations on the data, as specified by the instructions, can be carried out.
- The resulting data can be stored again.
- Communication with the "outside world" is provided.

A microprocessor chip generally cannot operate as a whole computer: a number of functions have to be added which are usually provided as additional integrated circuits. Since the microprocessor and those associated chips should be compatible, each microprocessor tends to expand into a large "family" of IC's.

Some examples of such associated IC's are:

- A memory containing the program, often in the form of a read-only memory, ROM.

- A memory for the data, usually a random-access memory, RAM.
- Input and output interfaces (I/O), for instance for the conversion of serial input data as from a keyboard to a parallel form acceptable by the microprocessor.
- A clock, normally employed to synchronize the various processes in the microprocessor.
- A power supply.

Usually the interaction between the microprocessor and the associated IC's takes place via buslines for addresses, data, and control instructions.

Most microprocessor manufacturers also deliver printed-circuit boards with an arrangement of microprocessor and associated IC's which can be used as a small computer (single-board computer).

For more complex systems a large number of IC's are available performing special functions:

- I/O controllers for various devices such as floppy discs or visual display units.
- Controllers to manage more complex memory configurations (including memory protection options).
- Processors for specific arithmetic functions, such as fast multiplications.
- Interfaces for handling analog output signals (e.g., for actuators).

In the past ten years of microprocessor development, two directions have become clear:

- Several of the associated IC's mentioned above have been included in the microprocessor IC, thus reducing the number of external chips needed and thus reducing the total system cost.
- The number of bits which are treated in parallel has increased. The first designs operated with 4-bit instructions and data. Soon the 8-bit microprocessors were developed, more recently followed by 16-bit designs. The first 32-bit microprocessor was brought to the market in 1980.

In this way much more powerful microprocessors have become available. At the present time the 4-bit variety is made in large volume and is applied in simple control systems and in toys and games. The 8-bit microprocessor, in market size the largest category, has been extensively used in countless applications and in widely varying degrees of complexity. The newer 16-bit machines are meeting with some restraint because users must make much larger investments in software.

INTEGRATED CIRCUITS, PRODUCTS

Software Aspects

The proliferation of microprocessor hardware has brought about large efforts in software development. These have been borne partly by the IC manufacturer and partly by the user.

The application of a microprocessor in any computerlike application requires at least some elementary software, as in supervising the operation of the system or in facilitating the use of a simple assembly language. With the growing complexity and power of the microprocessor, more such software is needed. For instance, if the new 16-bit microprocessors are to compete with the current line of minicomputers, they should have at least somewhat comparable levels of software packages.

Most microprocessor manufacturers have provided development systems, which allow the user to develop and test a program in some high-level language. Such a program is then translated into a form that fits the particular microprocessor. Advanced systems also provide the environment needed to test the operation of a program in a given microprocessor configuration (in-circuit emulation). Many of these development systems are designed for one specific brand of microprocessor, but more widely applicable systems have become available.

The user must write a program which fits a specific application. To a large extent this has been done in the assembly language available with the microprocessor: a brand-specific set of simple acronyms allowing a step-by-step description of the program to be executed. Though this arrangement allows the optimal use of available memory space and processing capabilities, programming is rather slow and inefficient. For longer programs in more complex systems, it is desirable to use a high-level language which permits a much more condensed form of writing instructions. There are obvious advantages:

- A much higher productivity of programmers can be obtained.
- The resulting programs are much more "portable"; a given program can run on various types of microprocessors.
- Documentation and maintenance of the software package are much easier because of the broader descriptive nature of the language.

Much attention has been paid to the definition of subsets of languages like Basic and Fortran suitable for microprocessor use and to the development of the necessary compilers. Recent favorites in high-level languages are Pascal and C-language, whereas the Department of Defense's real-time ADA language is expected to gain importance.

Some trends in the development of microprocessors with respect to bit length and on software are apparent. Currently the main market for microprocessors lies in the 8-bit types and the associated IC families. The introduction of a 16-bit type requires that the user make large and lasting commitments in software development. This will be done only in those cases where the advantage of doubling the word length is obvious. Perhaps the 32-bit versions will appear to be relatively attractive since they provide a very broad span of high-level technical applications.

The present course of the microprocessor industry has led to a collision with the minicomputer makers, who usually market 16-bit systems. Whereas the semiconductor companies have their strength in cheap manufacturing of large-series IC's, the minicomputer makers have great expertise in architecture and software, areas where currently much of the user's development money is spent. At the same time, semiconductor companies are attempting to broaden their systems and software expertise to compete with the minicomputer makers. On the other hand, the minicomputer manufacturers are building their own chip design and manufacturing facilities in order to maintain their independent competitive position. Thus in the 16-bit market the IC industry for the first time is encountering a strong competition with already established acknowledged levels of supporting software.

This situation will force the IC makers to develop a high level of systems software around their microprocessors. It is apparent that only a few companies will be able to make the necessary investments, and the 16-bit microprocessor market will probably be divided among them. On the other hand, minicomputer makers may well move toward more complex computer systems, thus exploiting their strength in systems architecture and software.

The new 32-bit design from Intel has the power of a medium-sized mainframe computer. Recently, IBM experimentally built a major part of an IBM-370 computer on three chips.(6) Both in the 16-bit and 32-bit markets the IC manufacturers are deeply interested in novel mass-market applications of their devices, through which they can obtain the usual benefits from the learning curve. It seems, however, that such applications will come rather slowly: there should be clear advantages over the existing 8-bit micros as well as over the existing minis or small mainframes.

As discussed earlier, the proliferation of software requirements is a major problem. Several solutions are possible:

INTEGRATED CIRCUITS, PRODUCTS   69

- Software development is being tackled as an engineering process instead of an individualistic "art" of programming. Specific procedures and tools are being developed and applied.
- Some microprocessor manufacturers have developed systems that accept the instruction repertoire of some favorite minicomputers, allowing existing minisoftware to be run on the micro.
- Introduction of high-level languages will considerably increase productivity of the programmer.

In some designs software is "put back" in the form of integrated circuits.(12) For instance, a mathematical software package usually resides in the program memory of the micro. An alternative is to build a "coprocessor" IC, in which the necessary programs are available in ROM. This coprocessor can also perform the mathematical calculation if ordered to do so by the microprocessor. This scheme decreases the load and software complexity of the microprocessor. The definition of suitable packages of such "firmware" has just begun. Examples are the use of a "chip compiler" for Pascal as developed by Intel(7) and the architecture of that firm's 32-bit micro, which is suitable for executing the new language ADA.(13)

GENERAL AND SPECIFIC DIGITAL LOGIC

An impressive array of digital circuits is available for a wide variety of applications. Within the scope of this discussion, it is impossible to give an overview of the entire field. It spans the spectrum from rather simple, broadly applicable IC's such as gates, registers and counters, timers, adders, and multipliers to more specific circuits such as code converters and output drivers.

In bipolar technology a number of "logic families" has been developed in which the various circuits are compatible regarding input/output and speed. They are based on different ways of constructing the basic logic gate, e.g,:

- Transistor-Transistor Logic of TTL, the broadly applied workhorse for digital technology.
- Schottky-TTL, a TTL version with higher speed, using so-called Schottky diodes.
- Emitter Coupled Logic or ECL, applied where high speed is essential.

In addition to these logic families, bipolar IC's have been designed with specific functions in several other topologies like I^2L or ISL when special requirements are present. Obviously, logic circuitry is also available in various MOS technologies, such as pMOS, nMOS, and CMOS. Their properties vary in terms of cost, possible circuit complexity, speed, and dissipation, depending on the technologies applied.

With increasing technological possibilities, circuits grow more complex (VLSI) and thus often more specific. This tends in principle to lower series because of the reduced application range. Thus, higher costs associated with increased complexity (design costs, high capital investment for manufacturing and testing) are not automatically offset by a larger turnover (as is the case in memory circuits). Much care has to be exercised to target the IC onto a specific application, which requires a good systems knowledge in the design phase. Examples are IC's for digital telephone, speech, synthesis, and computer peripherals.

Whereas memories and microprocessors are large-volume commodity items, the more specific types of digital integrated circuits will often sell in too small a volume to be profitable for the merchant house. As a result, many systems houses have started their own captive IC production.

Alternatively, a "custom IC" business is emerging: companies specializing in the design and manufacture of IC's on a specification by systems companies. Though in the past such activities never really substantially developed,(8) there may be a change when more sophisticated design tools become available.

UNCOMMITTED LOGIC ARRAYS

The Uncommitted Logic Array or Gate Array is an old idea, currently getting revived attention as the cost of IC design increases sharply with growing complexity.(9) It aims at a minimum design time, low cost, and fast delivery for small-quantity customized IC's.

Essentially, an Uncommitted Logic Array consists of an unfinished IC with a large number of separate unconnected transistors. The last process steps have not yet been carried out - in particular, the metallization step connecting the various transistors to form a circuit. The mask used in this step is made on the user's specification, thus transforming the general array to a specific circuit suited for an application.

We see that there are two essential ideas behind Uncommitted Logic Arrays:

- The basic chip with the unconnected transistors can be cheaply produced in large quantities and kept in store by the manufacturer.

- The specific adaptation to the user's requirements asks for only one mask design, which can be made relatively cheaply.

The advantages of Uncommitted Logic Arrays are a low cost in comparison with a fully customized design, and a relatively short time needed for design and manufacturing. Some disadvantages are the not optimal use of silicon area since a number of elements will not be used, the difficulty in designing efficient connections between the various elements, and the limited number of available pins.

The growing interest in Uncommitted Logic Arrays leads to new products and services. New combinations of transistors (sometimes prearranged as functional blocs) for digital and analog circuits are designed with an eye on specific user groups (requiring, of course, good insight into their systems requirements). Flexible computer-aided design support systems will be made available, preferably at the user's premises, to allow full use of the technology's potential.

Uncommitted Logic Arrays are available in several bipolar technologies and recently also in CMOS. Their ultimate market potential is difficult to predict. One might expect that they will constitute an interim technology for some time to come. Ultimately, true custom-design techniques should become so sophisticated that they will become more attractive than gate arrays. Note that there is ample room for new ideas in gate arrays that are compatible with average technology requirements. This condition and the manufacturer need for good customer relations make the field interesting for new venture businesses. However, well-established companies like Texas Instruments, Motorola, and Signetics are showing much interest in this market. Also, new business structures are emerging, like separate design and manufacturing ("silicon foundries") companies.

ANALOG IC'S

Whereas in digital IC's the transistor circuitry is operated in only two states (e.g., the transistor is "off" or "on"), in analog IC's the circuit is generally assumed to perform some linear operation on an incoming signal (for instance, proportionally amplifying it). This task is much more sophisticated than just switching between an "on" and "off" value, and hence a much stronger control over the properties of the transistors and other circuit elements is needed.

This consideration limits the choice of applicable technologies. The vast majority of analog circuits are made in bipolar technologies, which allow the design of transistors with well-specified operating characteristics. However, in recent years MOS technologies (and especially CMOS) are also finding a place. In particular, MOS-Field Effect Transistors are advantageous for low-noise, high-impedance input stages of amplifiers and the like.

The range in analog IC products is extensive. The following somewhat arbitrary choice illustrates a few major categories.

General-Purpose IC's

This category contains a wide variety of circuits, such as operational amplifiers, phase-locked loops, signal filters, voltage regulators, and voltage-frequency converters.

The designer of a measuring instrument or control circuit chooses IC's as building blocks with the desired properties, such as amplification characteristics, phase-frequency response, or temperature behavior. This circuit will usually consist of a combination of discrete components (transistors, diodes, resistors) and integrated circuits. The exact function of the IC is often determined by external circuitry. For instance, the (external) feedback loop designed for an operational amplifier-IC may determine its behavior in the total circuit, permitting the use of one type of IC for a multitude of applications.

Special-Purpose IC's

These are IC's which have been developed with a specific application in mind.

- Consumer audio IC's, such as power output stage, FM or AM tuner, Dolby noise reduction.
- Consumer video IC's, such as filters, IF amplifiers, tuners, which are currently mostly analog (though increasingly digital techniques are being introduced here(11)).
- IC's for telephones, either still fully analog (touch-tone senders, ringing circuits) or intended for future semi-digital operation in the exchange (codec, subscriber-line interface circuit).

These types of circuits are mass-fabricated to perform an exactly specified function in equipment. The choice of technology depends on the requirements to be met and on cost considerations.

INTEGRATED CIRCUITS, PRODUCTS

Interfaces

Conversion from analog to digital signals and vice versa is growing in importance as digital signal processing techniques and computer applications are proliferating. For instance, in communications technology signals are originally in analog form, such as sound or pictures. They are transformed into an analog electric signal via a microphone or TV camera, transmitted to the destination, and reconverted into an analog (sound or light) signal. However, it is often advantageous to transform analog signals into digital form before transmission takes place. Digital transmission can guarantee maximum freedom from added distortion or noise. After transmission, signals must be reconverted from digital to analog.

In all these cases analog signals have to be converted into digital information and often, further down in the system, into analog again. The digital signal should be an accurate representation of the analog signal. This requirement can be met by a proper choice of the frequency of sampling of the analog signal (how many times per second the voltage value of the analog signal is transformed into a digital value) and the number of bits needed for each sample (depending on the resolution and dynamic range needed).

For instance, a TV signal may be faithfully represented by 8-bit words, repeated 18×10^6 times per second and a Hi-fi audio signal by 14-bit words but repeated 4×10^4 times per second. However, readings of a sensor controlling a chemical process might be adequately represented by 8-bit, once-a-second digital words.

Many forms of Analog-to-Digital (AD) and Digital-to-Analog (DA) converters have been developed, both regarding circuit design and technology. Often, hybrid technologies are used, allowing compatibility at the "digital side" with the bipolar-TTL or MOS technology usually applied.

Trends

The advance of digital-signal processing techniques may lead to a decrease in the importance of analog circuits. However, analog methods are often simple, elegant, fast, and cheap compared with digital techniques. Moreover, with an increase in process possibilities (for instance, creating better resistor, capacitor, and inductor functions on a chip), the range of options for analog IC's will be extended. Consequently, although some of the market will be taken over by digital technology, it can be expected that an increasing world market for analog IC's will develop in the next decade. Typically, audio and video products in the lower price ranges will continue to rely on analog IC's for a long time.

MISCELLANEOUS DEVICES

A large number of devices have been developed or are being developed which use IC-like technologies which differ considerably from those we have just discussed. Regarding current business volume these devices are rather insignificant. However, since several developments may become important in the future, we briefly survey a few examples.

Silicon-on-Sapphire (SOS)

This technology, in which an epitaxial silicon layer is grown on sapphire, promises high speeds and low power. Military applications are interesting because of the apparent good radiation resistance. However, the cost of sapphire and the reportedly rather difficult technology have prohibited the introduction of SOS on major IC markets.

Gallium Arsenide (GaAs)

The physical properties of the semiconductor gallium arsenide suggest that it can be used to manufacture transistors that are inherently faster than those in silicon. Indeed, GaAs transistors have been developed for various high-frequency applications such as microwave amplifiers. Several fast digital integrated circuits have been made, though with relatively few components. The technology of GaAs is rather difficult and the technical progress slow.

For larger-scale digital integrated circuits GaAs may suffer from the drawback of a relatively high dissipation, which may limit the degree of integration. It is not clear how much improvement in speed will be obtained in practice with GaAs digital LSI. It is expected, however, that the technology will find only limited applications in areas where speed is a premium, such as microwave devices and front-end circuits of fast scalers, or fast computer circuits.

Charge-Coupled Devices

CCD's are MOS-like cells, arranged in pipeline configurations. Each cell is able to store an electrical charge and transfer it on command to a next cell.

Though various designs have been around for several years and a number of interesting applications exist (delay lines, memories, filters), CCD's have not yet gotten off the ground. They will probably find a more widespread use in

INTEGRATED CIRCUITS, PRODUCTS 75

light-sensitive devices and infrared detectors. A potentially very important application is a CCD memory for storing the picture (a frame) in a TV set. This would allow a substantial improvement of the picture quality. Here CCD's would find a first mass-market application.

Light-Sensitive Devices

Solid-state devices which transform light into electrical charge have been applied in several light-sensitive detector arrays. Various forms such as line arrays have found applications, for instance in photographic equipment and optical character readers. Solid-state replacements for television registration tubes have been developed for consumer TV cameras.

The light-sensitive devices are usually integrated with a type of CCD structure which transfers the information to the circuit's output. Sometimes, additional signal processing is also integrated on the chip. A sizable market for this type of device can be expected in the near future.

Magnetic Bubble Memories

Most semiconductor memories are volatile: when the voltage is switched off, the information disappears. Magnetic bubble memories do not have this disadvantage: as long as a permanent magnetic field is applied, the information remains stored.

The technique is based on the possibility of producing a very tiny magnetized area, the bubble, in a thin layer of magnetic material and moving this bubble from one discrete position to another along a ladderlike electrode structure. In this way, series of bubbles (representing "ones") and places where no bubbles exist (representing "zeros") can be put into a ringlike structure where they can be circulated (and thus "memorized") infinitely. The capacities of bubble memories are very large (on the order of millions of bits), but they are relatively expensive and need some peripheral electronics.

Bubble memories are now entering the market, but it is too early to discern where their major application will be. Quite recently, however, several U.S. companies have terminated their efforts in magnetic bubble memories. Apparently, market expectations are judged insufficient to justify the large investments needed to build up a production capability.

Josephson Junctions

Supercooled Josephson junctions are the fastest logic switches around. As was discussed in the section of Chapter 2 on computers, IBM has made a great effort in their development. Though the application of Josephson junctions in computers seems to be promising, many technological problems remain which delay their practical use in commercial equipment.

Thin-film Transistors

In a number of cases it would be useful to spread a transistor array over a large area. For instance, such an array with the size of a legal pad with a sufficient number of transistors could be used to drive matrix displays based on phenomena like gas discharges or liquid crystals.

However, single-crystal silicon wafers of this size on which transistors can be made are not available. Other semiconductor materials applied on glass substrates have been used to produce operating TFT arrays. Although the technological problems associated with large-volume production seem to be at odds with an early introduction, several possibilities exist for innovations in this potentially interesting and high-volume technology.

Opto-Electronic Devices

In this category we find the solid-state laser and avalanche photo detector used in connection with fiber-optical transmission techniques. Probably such devices will be integrated in the future with GaAs-like circuits to form efficient, cost-effective packages.

Another area is that of the use of light instead of electricity in digital and analog devices, which is made possible by IC techniques. This is still very much an area for basic research. An exception is the creation of an integrated optical wave-form analyzer for electronic-warfare applications.

Surface Acoustic Waves (SAW)

Several types of acoustic waves can be made to travel on the surface of thin layers of certain materials. With suitable electrode structures, applied on the surface of the material, surface waves can be generated and reflected to make tuned oscillators or filters. SAW devices have been introduced in such diverse equipment as military radio equipment and TV receivers. Their future application range may remain rather narrow, though the production volumes may be large.

INTEGRATED CIRCUITS, PRODUCTS

SOME FINAL NOTES

A dazzling array of IC products has been discussed in this chapter. There are several main trends.

- The driving forces in IC technology are found in the digital field. The computer industry in particular has been an eager purchaser of highly sophisticated digital IC's. But they have also been extensively applied in emerging new equipment like pocket calculators and in digitalization of what were analog products.
- MOS is the most widely used technology in digital IC's. It is applied in many mass-market IC's but also in expensive, very advanced memories and microprocessors. A variation, CMOS, is finding increasing use in view of its low dissipation. Note, however, the particular properties of bipolar IC's make them preferred in a wide range of more special applications.
- Perhaps the fastest technical advances are made in memories and microprocessors. The development of memories critically depends on very advanced manufacturing technologies. In contrast, microprocessors rely more on systems know-how and software abilities.
- Several forms of (semi-) custom IC's are gaining importance, in addition to the PROM and the microprocessor, which are already amenable to individual programming. Gate arrays quickly find wider applications, and the indications are that full-custom IC's will also gain importance.

Finally, note that the picture sketched above is that of the most important IC market, that of the U.S. The European and Japanese markets are distinctly different, and therefore their industrial production has a different emphasis.

REFERENCES

1. Hodges, D.A., "Microelectronic Memories," in Microelectronics, A Scientific American Book (San Francisco: W.H. Freeman & Co., 1977), p. 54.

2. Fullagar, D., "CMOS Comes of Age," IEEE Spectrum, December 1980, p. 24.

3. Posa, J.G., "Microcomputer Memories Get Smart," Electronics, July 17, 1980, p. 92.

4. Toong, H.M.D. "Microprocessors," in Microelectronics, A Scientific American book (San Francisco: W.H. Freeman & Co., 1977), p. 66.

5. Veronis, A., Microprocessors: Design and Application (Reston, VA: Reston Publishing Co., 1978).

6. Posa, J.G., (ed.), "The System 1370 Processor Chip: A Triumph for Automated Design," Electronics, Oct. 9, 1980, p. 139.

7. Bernhard, R., "Computers: Micros and Software," IEEE Spectrum, January, 1981, p. 38.

8. Wilson, Robert W.; Ashton, Peter K.; and Egan, Thomas, Innovation, Competition, and Government Policy in the Semiconductor Industry (Lexington, MA: Lexington Books, 1980).

9. Electronic News, "Semiconductor Firms Turn to Gate Arrays," April 20, 1981, p. 50.

10. Posa, J.G., "Memories", Electronics, Oct. 20, 1981, p. 130.

11. Fischer, T., "Digital VLSI Breeds Next-Generation TV Receiver," Electronics, Aug. 11, 1981, p. 97.

12. Bernhard, R., "More Hardware Means Less Software," IEEE Spectrum, December 1981, p. 30.

13. Rattner, J., and Latkin, W.W., "ADA Determines Architecture of 32-bit Microprocessor," Electronics, Feb. 24, 1981, p. 119.

6
The Integrated Circuit Industry

In the preceding two chapters, IC's were reviewed in terms of technologies and products. We can now proceed with a discussion of the manufacturing industry: the companies that have not only used the opportunities provided by the new technologies but have also contributed significantly themselves to the development of new technologies and products. In this process the electronic industry has fundamentally changed. Many of the older, established electronic companies that were leading in the vacuum-tube era have not been able to keep this position and have either declined or concentrated on special markets. Many new ventures or relatively unknown companies have seized emerging opportunities and taken a commanding position. This has caused a substantial restructuring of the U.S. electronics industry in the past decades, the end of which is probably not yet in sight.

Though the industry seems to be stabilizing somewhat after a decade of great turmoil, it remains prone to further change. Technological change continues to exert its influence. For instance, the emergence of VLSI gives rise to opportunities for new ventures, but will be also change the traditional relations between component manufacturers and systems houses.

Manufacturing and selling IC's has become an extremely important business, conducted on a global scale. This is true regarding the sheer production value, which exceeds the $10 billion mark, and also with respect to the more strategic aspects of these components. Not only is their importance for national security relevant. The "basic capability" character of this industry for a country's industrial well-being makes the IC business a likely target for public concern.

The major production centers for IC's are the U.S., Western Europe, and Japan,(1) while some other Far Eastern countries are making rapid advances.(25) The U.S. industry has spearheaded the industrial development and application of IC's. Companies that specialized earlier in semiconductors, like Texas Instruments and Motorola, and newer ventures like Intel and Mostek established a world leadership in this new technology. The technology was driven by bipolar and later by MOS digital circuits which found important initial markets in the U.S. defense and space programs. Later on, the internationally dominant U.S. computer industry became a major customer for ever-improving circuits.

In addition to these so-called merchant houses - producers for the open international market - several large systems houses began their own "captive" manufacturing. Examples of early captive activities are IBM, AT&T, and Hewlett-Packard.

In the course of the years, changes can be observed in this structure of the U.S. industry: innovative, independent merchant houses and a few large companies with captive manufacturing. Several of the specialist IC companies diversified into systems, many were acquired by larger industrial entities, and captive manufacturing spread through the electronics systems industry. These changes have been brought about by economic necessity, but certain technical reasons can also be mentioned. At any rate, the basic structure of the industry differs considerably from that of a decade ago.

The dominance of the IC field by U.S. industry has been eyed calculatingly by European and Japanese governments. Several European countries have tried to establish their own IC industry protected by common European Economic Community tariff barriers. Sometimes this was done in the framework of their existing large electronics companies, at other times by forming new ventures. Japan has mounted impressively coherent efforts in electronics and integrated circuits, stimulated by a variety of government measures. These encompassed the consumer-electronics, computer, and telecommunication industry, and several of these companies have emerged as important IC manufacturers.

These considerations are the starting point for a more detailed discussion of the international IC industry, with the main focus on the U.S.

IC'S: A WORLDWIDE BUSINESS

The 1980 world production of the merchant industry may be estimated to be around $9 billion.(1) The U.S. accounts for nearly 60 percent of this world production. The Japanese industry is quickly expanding and its production value exceeds that of Western Europe, where an appreciable part is derived from subsidiaries of U.S. companies.

However, production statistics must be regarded with some care. In addition to the problem of accounting for American companies producing in Europe (or for European companies carrying out part of their manufacturing process in the Far East), a differentiation between merchant and captive production is sometimes difficult to make. Though in the U.S. a fairly clear distinction generally exists between the two types of production, in Europe and Japan most producers partly work for the open market and partly operate as captive producers. Usually, total production values are not published by the bigger companies and certainly not values for internal deliveries. Also, distinction between IC's and discrete semiconductors are not always made.

With regard to the consumption of IC's, the U.S. is again responsible for about 60 percent of world consumption, with Europe and Japan as approximately equal consumers.

The U.S. merchant production is divided among various product areas, as we discussed earlier (see fig. 5.1). The emphasis is strongly on digital IC's, with an important role for MOS technology. In contrast, Europe's production strengths are largely in bipolar analog IC's for consumer applications. Japan is also a large producer of analog IC's for its large consumer industry. Additionally, Japanese firms have developed a high-level MOS technology for memories during the last decade. In this cutting-edge technology they have become strong competition to the U.S. industry and now pose a distinct threat for its long-term prospects. They are also moving towards a strong position in the upcoming CMOS technology.

Estimates of the world markets for the different IC categories in the next few years predict a strong growth for MOS memory and microprocessors (between 20 and 30 percent annually). The market for digital bipolar circuits is also expected to grow, but at a slightly slower rate. The growth predicted for analog circuits is much less pronounced.

Such predictions must be modified by general economic trends. The economic recession that began in 1981 has already led to substantially modified predictions in the sales of several types of circuit. Additionally, an overcapacity has developed in the industry, resulting in fierce competition and drastically reduced prices, in particular but not only in the memory business.

As shown in table 6.1, the end users(1) of semiconductors in the U.S. are predominantly computer companies, whereas in Europe the consumer category is the largest.

Several studies have been devoted to the U.S. semiconductor industry in a world perspective. Some major reports have been prepared by the U.S. Department of Commerce,(2) the Federal and International Trade Commissions,(3,27) and Wilson, Ashton, and Egan.(4) Some useful data on the in-

Table 6.1. End Users of Semiconductors (1978).

	U.S.	Western Europe
Computers	56%	20%
Consumer, automotive	11	30
Industrial	11	18
Communications	9	14
Government	13	5
Miscellaneous	-	13
Total	100%	100%

Source: G. Dosi.(1)

dustry can also be found in publications of the Semiconductor Industry Association.(5,6) Several Congressional hearings have also made interesting material available.(7,8)

THE U.S. MERCHANT INDUSTRY

Introduction

The American merchant industry is presently dominated by a relatively small number of firms, most of which are located in California's "Silicon Valley." However, two of the leading firms, Motorola and Texas Instruments, have their headquarters elsewhere, at Phoenix and Dallas respectively. The older firms such as Fairchild Camera and Instruments, Texas Instruments, and Motorola have been firmly established in the semiconductor business since the 1950's. Many of the more recent entrants, such as Intel (1968), National Semiconductor (1959), and Mostek (1969) are spin-offs of earlier firms. Motorola, Texas Instruments, Fairchild, Intel, National Semiconductor, Mostek, Advanced Micro Devices, and Signetics presently account for around 75 percent of the American production value of integrated circuits (see fig. 6.1), excluding the U.S. captive industry.

Some of these newer firms which were established in the late 1960's have enjoyed impressive growth in sales during the decade of the 1970's. This growth has been particularly

INTEGRATED CIRCUIT INDUSTRY   83

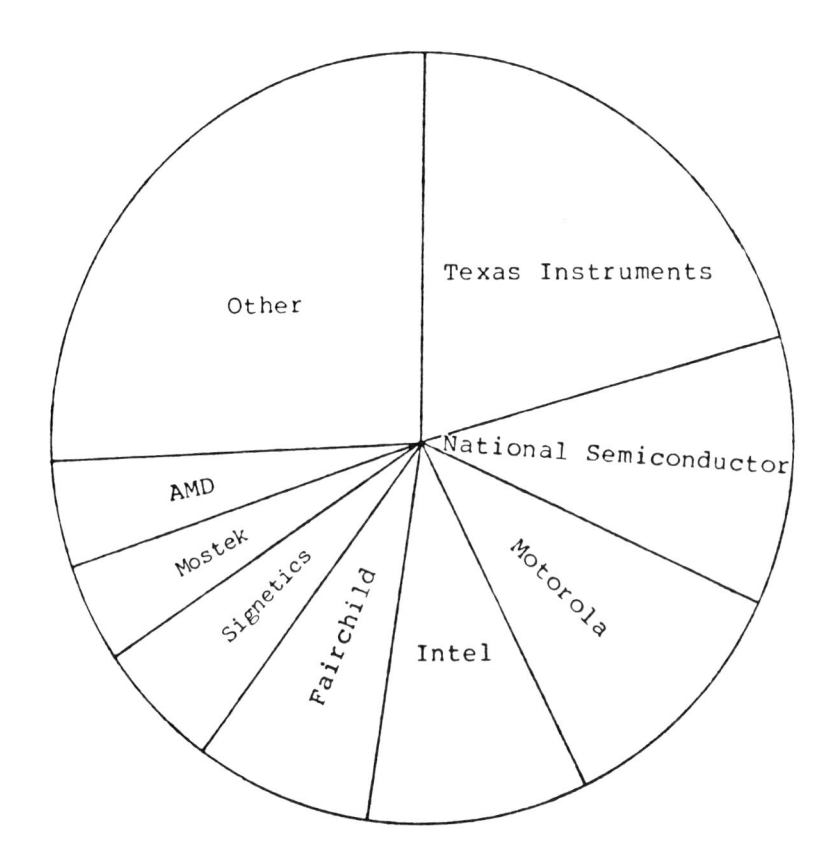

Fig. 6.1. 1979 IC production value of U.S. companies.

Source: Philips estimates.

pronounced during the last few years (1977-80). Several of these companies are now approaching the billion-dollar level in sales.

In the last decade there have been relatively few new entrants into the industry. This can be explained in several ways. In part it may have been caused by the lack of venture capital due to the earlier high capital-gains tax. Another factor has probably been the substantial capital requirements necessary for manufacturing today's complex chips. Recently, as revised tax measures led to a replenished flow of venture capital, several new ventures have been spawned. They often act as custom chip producers. An example is LSI Logic Corporation, a manufacturer of gate arrays, founded in early 1981 by Wilfred Corrigan, former president of Fairchild. Firms like this develop a market-niche strategy in the industry.

Overall, the industry has changed dramatically in the last decade. The relative position of some of the top firms has changed considerably; several have lost their market share to newer firms such as Intel, Mostek, and Advanced Micro Devices (AMD). With the exception of a few of the older, more established firms, the industry consisted of independent entrepreneurial companies in the beginning of the decade. Since 1975 there has been a trend toward greater concentration, including mergers and acquisitions.

In recent years many of the Silicon Valley firms have been the targets of takeovers by major corporations, including many European firms. Some examples include Fairchild, acquired by Schlumberger in 1979, and Signetics, which was acquired by N.V. Philips in 1975. Mostek Corporation was acquired by United Technologies Corporation in 1980 and Intersil by General Electric in 1981. The present ownership of the merchant industry is reviewed in table 6.2.

Many see these changes as having important implications for the industry in the future.(35) Some observers believe that firms that now form part of larger companies will undertake a shift from a technological, innovative focus to a more traditional business orientation. On the other hand, the large multinational companies which have acquired Silicon Valley firms may be in a better position to absorb temporary losses while pursuing long-term goals of increasing market share and product positioning. In particular, this advantage may permit the acquired firms to undertake more extensive research and development, as in the corporate research laboratory that Philips established at Signetics.

In the past the industry has been very sensitive to recessions, as illustrated by the industry's cutback in capital expenditures and personnel cost during the recession of 1974-75. It is widely believed that as a consequence the industry was not able to meet rapidly rising demand in subsequent years, allowing Japanese firms to gain a larger market share.

The learning curve is fundamentally important in this industry. As explained in Chapter 2, the cost of manufacturing of a particular product predictably decreases when the total volume produced increases. Forward pricing has become a common strategy. The price of a new product is lowered early in its life span, aiming to achieve a large market share and large-volume production. Later on, when a strong market position is attained, prices are adjusted to a profitable level. However, substantial capital expenditures in capital goods and R&D are required in such a forward-pricing strategy in anticipation of future increases in sales.

Table 6.2. Acquisitions in the U.S. Semiconductor Industry

Company In Which Participation Is Taken	Participating Company	Country	% of Equity Owned	Year of Acquis.
AMD	Siemens	Germany	20	1977
AMI	Gould	U.S.	100	1981
AMS				
Analog Devices	General Electric	U.S.	100	1980
Electronic Arrays	S.O. Ohio	U.S.	20	1977
Fairchild	N.E.C.	Japan	100	1978
Interdesign	Schlumberger	France/U.S.	100	1979
Interset	Ferranti	U.K.	100	1977
Litronix	General Electric	U.S.	100	1980
Mos Technology	Siemens	Germany	100	1977
	Commodore	U.S.	100	–
MOSTEK	United Technologies	U.S.	100	1979
SEMI	GTE	U.S.	100	–
SPI	CIT-Alcatel	France	25	1981
Signetics	Philips	Netherlands	100	1975
Siliconix	Lucas	U.K.	24	1977
Solid State Scientific	VDO	Germany	25	1977
Spectronics	Honeywell	U.S.	100	–
SSS	Thomson	France	100	–
Synertek	Honeywell	U.S.	100	–
Zilog	Exxon	U.S.	50	1976

The keen price competition existing in the industry is also caused by the similarity in products offered by various firms. High-volume producers of standard components such as Texas Instruments and National Semiconductor have been particularly noted for their aggressive price cuts in a forward pricing strategy. Others like Intel, however, try to be the first in a new market with innovative higher-priced products and leave the market when price erosion sets in.

During the last decade a significant amount of forward and backward integration has taken place in this industry. A number of the larger firms such as Texas Instruments, Motorola, and Fairchild have integrated backward into purifying silicon, growing silicon crystals, and making silicon wafers. Additionally many firms have practiced forward integration with varying degrees of success. Examples include integrated-circuit firms entering new markets, such as digital watches, pocket calculators, and computers. However, these product lines are sold in a different market from integrated circuits and semiconductors. Products such as watches and calculators require a specific type of marketing, which is different from the firm's usual skills. Some companies have been successful diversifiers, notably Texas Instruments, while others like National Semiconductor have experienced several setbacks.

Texas Instruments

The largest firm in the U.S. industry, Texas Instruments (TI), accounts for more than 20 percent of American noncaptive production of integrated circuits. In earlier days the firm was a major recipient of military contracts. It is now a major participant in the Very High Speed Integrated Circuit (VHSIC) program for the Department of Defense.

Texas Instruments is a highly efficient producer and is known for its marketing strategy of undercutting competitors' prices to capture a greater market share. In fact, it was the first in the industry to apply this concept. The firm commands a wide range of technologies and products. It has a notable track record in diversification with new products in a wide range of consumer and professional applications. Its management structure systematically promotes innovations (through the Objectives, Strategies and Tactics programs).(9)

The firm is giving increasing importance to technological leadership and product innovation.(28,29) It has surpassed its sales targets, achieving sales of over $4 billion in 1980, and has revised its sales goal for 1990 from $10 billion to $15 billion.(10) Some growth problems have occurred since.(36)

Motorola

One of the older firms in the industry, originally active primarily in mobile radio equipment, Motorola has recently shifted its emphasis from IC's for consumer electronics to the more rapidly expanding markets in data communications and automotive electronics. In several areas the company has obtained a leading technological position, for instance, in RAM's. The firm's total sales in 1979 were $2.7 billion, of which nearly $1 billion was in semiconductors.

Motorola also took the road to further diversification. The firm acquired in 1981 the computer maker Four-Phase, which is being integrated with Motorola's existing data-communication division.

National Semiconductor

National Semiconductor is the second-largest integrated-circuits producer in the U.S. and is rapidly expanding.(11) In the last few years sales have increased substantially, outpacing the rest of the industry. The firm has practiced both forward and backward integration. Recently National acquired a computer business from Itel, and it markets computers in cooperation with Hitachi. Still, about 80 percent of its turnover is in semiconductors. The firm has recently been eliminating some of its less profitable products such as calculators, watches, and magnetic bubble memories.

National has an orientation toward consumer electronics and is a leader in linear circuits. Although sales have increased rapidly in the last few years, profit margins are low. However, the company runs a very cost-effective operation in which efficient manufacturing is critical. The firm's sales revenues in 1980 were $980 million.

Intel(12,13)

One of the newer firms in the industry, Intel was founded in 1968. In 1971 the firm introduced the microprocessor. The firm has always been on the leading edge of technologies and products. Intel benefited in the earliest years from its systems orientation, a trend which still continues as its microprocessor products become more and more complex. At that time the major semiconductor companies were oriented toward electronic components rather than systems. Intel now spends about 10 to 11 percent of its annual sales on R&D.(14) The firm is strong in MOS technology and has the greatest percentage of products manufactured for the EDP area of any firm in the industry.

Intel has developed a rather different pricing strategy in the past. The firm has used a "cream skimming" strategy rather than a forward pricing strategy used by a company such as TI. Intel usually introduces a product early but at a high price and withdraws the product from the market after other firms begin marketing the product at lower prices. This market skimming strategy has been oriented toward the military and data-processing markets.

In producing microcomputers, memory boards, and development systems, the firm has successfully practiced in-house forward integration. Only the company's digital-watch product line proved to be a failure: the firm lacked the needed marketing expertise and withdrew the line in 1977.

With respect to its expanding microprocessor line, Intel is emphasizing systems design and software and has acquired a software house (MRI Systems, 1978). The firm is developing standard software, such as operating systems and high-level languages, which it plans to produce and market in the future. By developing new software the company hopes to reduce much of the programming costs to its customers and thus improve the marketability of its products. The firm has achieved impressive sales growth over the last decade; in 1980, sales were $854 million.

Fairchild Camera and Instrument

Fairchild is one of the older firms in the industry and has been the leading semiconductor company for many years. However, quite a few of its executives and engineers left the firm and started their own companies (the "Fairchildren," such as Intel, in 1968). The company has had a relatively weak performance during the last decade. It was acquired by Schlumberger in 1979 and a new, long-term management philosophy has been implemented. The firm is strong in ECL and bipolar devices, but has neglected MOS in the past.

Mostek

Mostek, founded by former employees of Texas Instruments, is strong in MOS devices. The firm initially achieved a good market position with n-MOS memory devices thanks to an innovative design, at first with the 4K RAM chip and more recently with its 16K RAM chip. The firm's total semiconductor sales were estimated to be around $230 million in 1979. The firm presently has a strong orientation toward producing components for the communications market. In 1980 it was acquired by United Technologies.

Signetics

Signetics, now owned by the U.S. Philips Trust, has a product line which is entirely composed of IC's. The firm is strong in bipolar logic and memory devices. It also produces MOS microprocessors and memories. A relatively large share of the firm's product line, now over 17 percent, is being manufactured for the government and military market. The company has access to Philips' European R&D in semiconductor technology.

Advanced Micro Devices

AMD was established in 1969 and Siemens acquired partial equity ownership in 1977. The firm has experienced rapid growth in sales in the last few years. Its entry strategy was to second-source IC's, in both bipolar and MOS technologies, from other firms in the industry. AMD then improved on the performance, reliability, and quality of these chips in its product line. As the firm's sales revenues grew, AMD changed its strategy to develop its own devices.

The firm is presently spending more on R&D than any other firm in the industry, about 12.5 percent of turnover. A combination of high R&D costs along with substantial capital expenditures has resulted in lower earnings. AMD has a high overhead, attributable to a push for a greater market share in the future.

A large part of AMD's production, about 60 percent, is for the EDP market. The firm also manufactures more than over 18 percent of its sales for the governmental and military market. The firm is not very interested in production of IC's for the consumer electronics sector. AMD is presently increasing its market share in MOS microprocessors.

AMD has been characterized as being one of the few firms in the industry that are driven more by the market than by technology, and it has experienced rapid growth. In 1980, sales were $225 million, compared with sales of $148 million in 1979.

THE CAPTIVE INDUSTRY

Major American electronic systems companies usually have a captive production of integrated circuits, for instance:

- Computers: IBM, CDC, Honeywell, NCR, Burroughs, Sperry-Univac, DEC, Data General.
- Communications: Western Electric, GTE.

- Instruments: Hewlett-Packard, Tektronix.
- Automotive: General Motors (Delco).
- Control: Honeywell.
- General: General Electric, RCA.

The total value of this captive production is hard to assess. Accurate production data are not available and estimating equivalent sales prices for such products is rather arbitrary. Industry estimates put the 1980 U.S. captive production of IC's at about one third of that by the merchant houses. Figure 6.2 shows the estimated production value of the major suppliers.

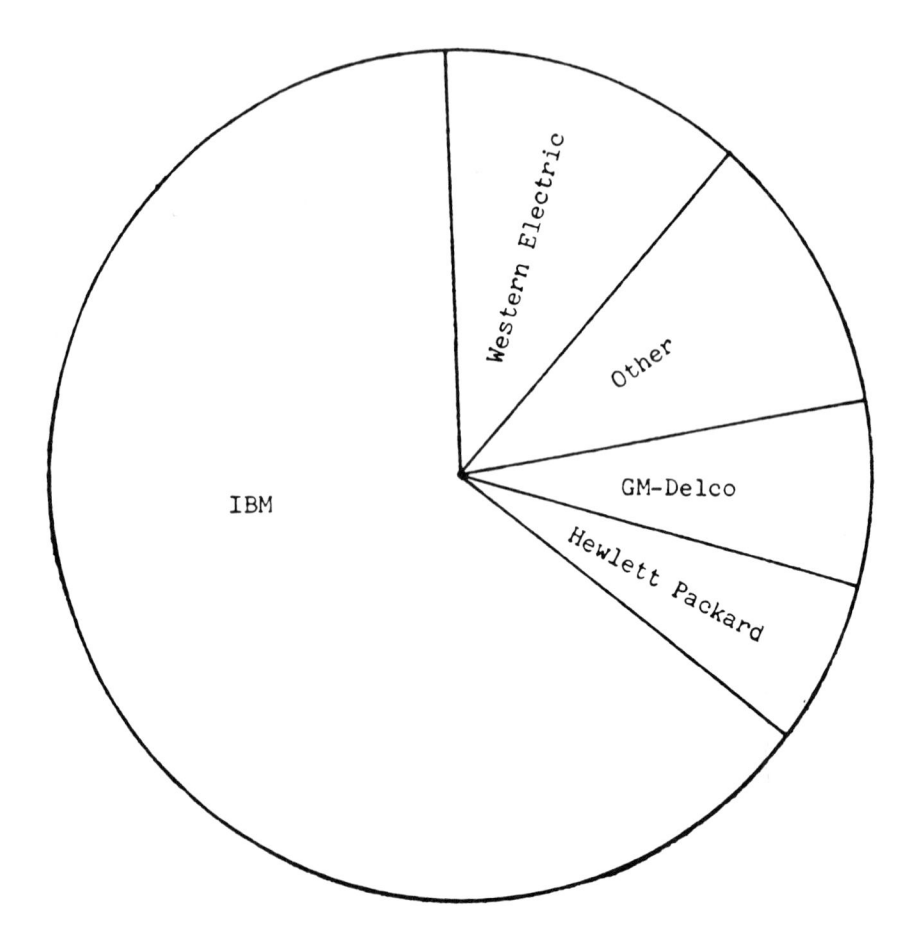

Fig. 6.2. <u>1980 U.S. captive IC production (in value)</u>.

Source: Philips estimates.

As Bloch(15) has indicated, an internal production of integrated circuits is governed by a set of requirements that differs distinctly from that of the merchant industry. Let us discuss some of his observations.

From a technical perspective, systems houses require that the system of which the IC is a part must be optimized, instead of the chip itself. This leads to a focus on compartmentalization and packaging: how can IC's be combined into an optimized "box" or subsystem? Special technologies have been developed to achieve optimal packaging. Also, an extraordinary emphasis is placed on quality and reliability, embedded in designs as well as in manufacturing processes. Finally, the production capability for a certain IC must often be sustained during the system's lifetime. This requires an exceptional stability of such a production process over many years.

Furthermore, from a management perspective the captive supplier acts differently. The point here is that an integrated company allows optimal interaction between systems specialists, circuit designers, and the technologists. Therefore, very advanced solutions are more likely to occur internally in a systems house than when a design is made with an outside supplier. In addition, the merchant industry's emphasis on cost of the IC is replaced by a major concern for a timely supply of the needed quantities and their quality and reliability.

Other arguments in favor of a captive production may include the wish to keep proprietary knowledge as much as possible inside the company and the reluctance to see a major portion of a company's "added value" going to suppliers.

One should note, however, that the cost of setting up and operating a captive production facility is appreciable and at times the facilities are underutilized. This has prompted some companies also to engage in the merchant market (e.g., recently NCR). However, it seems doubtful whether such moves will be very successful, in view of the dual commitment involved.

IBM

The largest data-processing equipment manufacturer in the U.S., International Business Machines, is also by far the largest captive IC producer in the United States. In fact, its IC production substantially exceeds that of the largest merchant company. Yet some of IBM's IC requirements are supplied from outside. Because of its sheer size, IBM strongly influences IC market conditions.

The integrated-circuit operations are fully centralized in the General Technology Division, which is a part of the Data

Processing Division. Major production facilities are located in East Fishkill, New York, and Burlington, Vermont.

IBM commands a range of technologies in bipolar as well as in MOS. Gate arrays are widely used. Considerable advanced technology is developed in-house, including electron-beam machines, ceramic substrates with multilayer connections, methods for connecting chips to such substrates, and a highly automated production technology. In addition to a high-level factory automation, this company probably has the most sophisticated CAD systems in the industry.

A strong research and development activity supports the fabrication sector. Here also, several less prominent fields are studied and developed. Notable areas of activity include Josephson junction switches and bubble memories.

Western Electric

Western Electric is the manufacturing arm of the American Telephone & Telegraph Company. It produces a part of its own need for telecommunications-related integrated circuits, though it still purchases a sizable fraction on the merchant market. It tries to concentrate its own production on chips of which the proprietary nature provides a competitive advantage.

The technological basis for Western Electric is provided by Bell Laboratories, perhaps the world's most prestigious industrial R&D establishment. It has frequently introduced new developments in semiconductor technology. Currently a second-generation electron-beam machine is being developed and a complete method for X-ray lithography has been announced. Some technologies have been licensed to other companies such as E-beam patents to Varian.

Bell Laboratories is responsible not only for basic research but also for development of the processes and design of integrated circuits to be used by Western Electric. A recent example is the development of a 32-bit microprocessor. The company has a substantial capacity in circuit design systems and a large pilot factory for the production of IC's.

Western Electric's major production facilities are located in Reading and Allentown, Pennsylvania. It has a range of technologies available, bipolar as well as MOS.

The strong tendency toward digitalization in communications networks will lead to an accelerated need for integrated circuits. This will probably result in a strong increase in Western Electric's captive activities.

General Motors - Delco

General Motors is the only American automotive industry with a captive IC production facility. It has concentrated its integrated-circuit production at Delco. This company has developed an nMOS technology and it licensed microprocessor technology from Motorola.

Though GM is already a major producer of IC's, much of their requirement is bought from the merchant houses. Delco's production capacity is said not to have been increased in proportion to the demand for IC's because in recent years large investments had to be diverted to retooling of the car-manufacturing operations.

Honeywell

Honeywell's captive activities are concentrated in the Solid State Electronics Division (SSE). Its headquarters are located in Minneapolis and the wafer fabrication is located in Colorado Springs. Mounting is carried out in Mexico and the Philippines. Honeywell is also active as a merchant house in nMOS and CMOS through its Synertek acquisition, in optoelectronics (Spectronics), and in infrared devices for military applications.

The firm's captive production is necessary to ensure the production of IC's that are optimized for the systems in which they are going to be used. The devices are often produced in small quantities, although there is also high-volume production of Hall sensors for the automotive market. They usually contain a considerable amount of special knowledge that cannot be copied easily by competitors. New types of circuitry often give opportunities to construct new systems with improved performance or reduced cost.

Honeywell is a specialist in the field of integrated sensors, to which much R&D has been devoted over the years. Signal-conditioning and signal-processing circuitry are often integrated on the chip. Examples are integrated transducers for "conventional" measurements such as pressure, temperature, humidity, and magnetic induction. An optical image sensor with CCD signal processing, to be applied in single-lens reflex photographic cameras, may open up new fields of business.

A wide range of technologies is needed, from MOS and bipolar to the special processes needed to manufacture the sensors. Quality and reliability are better than in the average merchant house because considerable attention is focused on that aspect, in part required by the military specifications of many products. In addition, packaging often receives special concern. Testing is another field of expertise, certainly in relation to the sensors.

Circuit design is carried out in the product divisions, where currently nine design centers exist, or in the SSE Division itself. The design methodology is centrally chosen and imposed throughout the company.

The SSE Division, reporting to the Vice President for Development, is partially funded from corporate sources. These funds are used for general development of technologies, their documentation, and so forth. The recently obtained VHSIC contract will strengthen the technology for submicron devices.

Hewlett-Packard

Hewlett-Packard has initiated captive production to maintain its line of top-quality measuring instruments and computers. To maintain a competitive edge, the end products must contain several critical IC's of a proprietary design. The special properties of the IC's may include performance (speed, noise behavior), circuit configuration, or quality. Captive production amounts to about 30 percent of all IC's used by the firm.

Hewlett-Packard is unique in that it has allowed each of its production divisions (currently forty-four) to decide whether or not to establish its own IC manufacturing operation. This practice has led to a decentralized pattern of nine production centers. They are supported by three research centers at the Corporate Research Laboratories. The programs of the Corporate Research Laboratories are strongly determined by future needs envisioned by the product divisions.

The technologies reflect the wide range of high-technology end equipment in which Hewlett-Packard is active. Included are digital and analog bipolar, nMOS and CMOS, SOS, logic and microwave GaAs, and optoelectronics. Obviously, the decision to put the IC production as close as possible to instrument design and manufacturing was prompted by the desire to get the best coupling and turnaround time possible.

In recent years much work has been devoted to the problems of VLSI design and technology. Hewlett-Packard has reported the attainment of an IC with probably the 1981 world record regarding component count.(30) In its VLSI design philosophy, basic strategies such as those proposed by Mead and Conway (see page 164, chapter 7) have been adopted. Necessary new equipment for a number of special technologies must be made in-house. A recent example is improved E-beam equipment.(16) Computerized process technology control is one of the methods developed to cater to the large variation in applications needed.

Hewlett-Packard will probably not continue to follow this policy of scattering its facilities, in view of the rapidly rising capital costs involved. It seems likely that its production efforts will be concentrated in a few places. Expertise in design techniques and communication technology available at Hewlett-Packard will allow design activities to be placed very close to its instrument design and manufacturing operations.

General Electric

Traditionally a manufacturer of electron tubes and later of semiconductors, General Electric withdrew from the integrated-circuit business roughly a decade ago when the company sold its computer operations to Honeywell. The technology, however, was further developed at the Corporate R&D laboratories at Schenectady, New York.

In the last decade this company has been changing considerably. Several new activities such as the production of plastics have been set up successfully and older ones were terminated during a period of rigid financially oriented management. Recently, however, top management defined a renewed focus on high technology. In response, several actions were taken to acquire in-house capabilities in microelectronics, which is increasingly replacing the electromechanical controls in many of the company's products, as in equipment for factory automation.

The existing basis in microelectronics at the Schenectady laboratories was used to establish large captive design and production facilities in North Carolina, at Triangle Research Park. They are intended to support the technical sector in General Electric with proprietary IC's. A second facility in Charlotte, North Carolina, is more directly oriented toward applications for industrial electronics.

In 1981 the California company Intersil was acquired. This relatively small company is known for its expertise in CMOS and power-MOS. Intersil is likely to act as a donor for manufacturing technology know-how to General Electric but will be run as a relatively independent semiconductor company.

In another acquisition, Calma was added to General Electric. This company is one of the leaders in computer-aided design (CAD) technology for electronics as well as mechanical design, an expertise that complements General Electric's capabilities. Conversely, the broadening of the financial basis may provide Calma with the needed capital to invest in the development of the very complex VLSI design equipment of the future. It will be interesting to observe how rapidly General Electric can get its renewed microelectronics activities off the ground and how it will handle the integration of its newly acquired facilities.

SUPPORT INDUSTRY

The semiconductor industry depends on a broad range of supporting industries: materials, processing equipment, design systems. Little research has been carried out into the operation of this industry.(24) In fact, many of the existing studies of the semiconductor industry ignore this sector. However, several significant business developments are taking place. For instance, the numerous, usually small companies are increasingly being confronted with high costs of developing and producing highly complex equipment needed for new VLSI technologies. It will be interesting to observe how the industry will acquire the financial means and technical knowledge needed.

Materials

Pure silicon for integrated-circuit fabrication is manufactured in the U.S. by Hemlock (Dow Corning), Monsanto, Texas Instruments, Motorola, and Great Western (GE). Major foreign suppliers include Wacker and Smiel (West Germany) and Osaka, Shin-Etsu, and Koma in Japan.

Generally speaking, the supply of silicon-wafer material is in the hands of specific materials companies. Many integrated circuit manufacturers have their own production to cover some of their individual needs. Additionally, it provides a useful capability for R&D purposes.

The diversity of materials needed by the semiconductor industry is supplied by a large number of chemical companies.

Equipment

Optical lithographic equipment is offered by companies such as Perkin-Elmer, GCA, Eaton, and Applied Materials (Cobilt). The Japanese company Canon is a contender here. Electron-beam machines for mask generation are produced for instance by Varian and Perkin-Elmer.

Processing equipment, including ion implanters, comes from Applied Materials, Eaton, GCA, Perkin-Elmer, Varian, and Veeco. Bonding machines are made by Applied Materials, Kulicke & Soffa, and Lindberg-Tempers.

Circuit design equipment is offered by Computer Vision, Calma, and Applicon. Testers are supplied by Fairchild, Gen Rad, Teradyne, and Tektronix.

This industry consists of a large number of more or less specialist companies. They entered the field either as new ventures or from a background in optics, mechanics, vacuum technology, or measurement techniques. Recently some efforts toward concentration have become noticeable: some companies are expected to acquire the capability to offer complete wafer-fab lines.

The U.S. equipment suppliers have achieved a leading position in the world, probably because of the easy transfer of people and technology that is commonplace in the semiconductor industry. A strong base of expertise has also been established in the major industrial laboratories of Bell Laboratories and IBM. Both, for instance, have developed E-beam equipment for in-house use.

Other Capital Goods

Many other capital goods such as clean-room installations are offered by the U.S. industry. The IC companies usually make a choice of some key components and develop their own fabrication line around it. Some new ventures sell equipment to connect different brands of equipment in a production line. However, aspects such as mechanization and computer control often enter as specific know-how of the IC manufacturer.

Complete processes are as yet not offered. The expertise to develop processes which give a company a competitive edge is usually found in the IC industry, not in the support industry.

THE EUROPEAN INDUSTRY

The European microelectronics industry is noticeably different from the American. It is characterized by a number of large vertically integrated electronics firms such as Philips and Siemens. In the past such firms have produced electron tubes and later semiconductor devices and have continued with IC's. Traditionally, they have been more oriented toward IC's for consumer electronics and telecommunications than those for data processing, since the European computer industries have remained relatively weak. In particular, Western European firms have been lagging in such areas as MOS memory and microprocessors. A large part of the European market is dominated by the American merchant houses. Figure 6.3 shows the 1980 market shares of the major companies.

A major strategy developed by European firms in attempting to acquire the most advanced technology is the practice of buying into American microelectronics firms. Examples of this include Philips, which acquired Signetics in 1975, and Schlumberger, which acquired Fairchild in 1979. Other firms

98   U.S. MICROELECTRONICS INDUSTRY

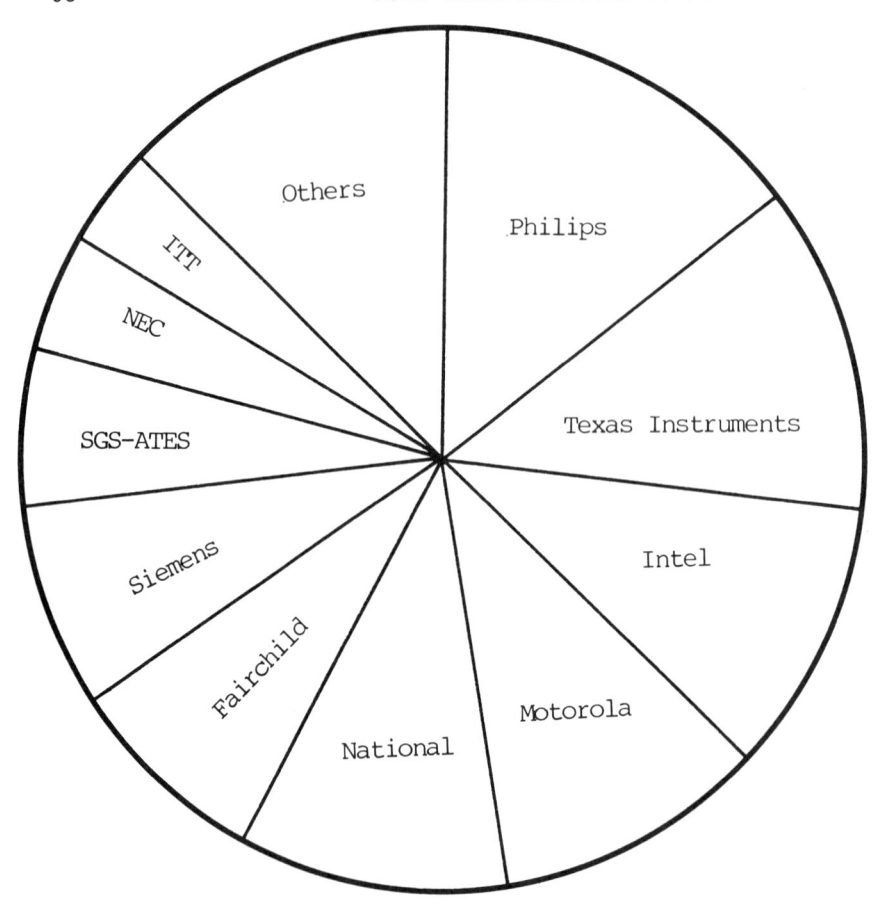

Fig. 6.3. European IC market (sales value).

Source: K. Jones.(22)

have purchased partial equity ownership in American firms. Examples here include Siemens, which has about 20 percent equity ownership in Advanced Micro Devices, and Robert Bosch, which owns about 25 percent of stock in American Microsystems. An alternative strategy is followed by some French firms which attempt to license U.S. technology in return for market access in France via joint ventures.

In view of this background, it is not surprising that European governments have undertaken microelectronics support programs to compensate for certain weaknesses in their industry.(17) This weakness may be illustrated in the industry's inability to meet European market needs. It supplies only 30 percent of the IC's sold in Europe, whereas American companies have nearly 60 percent. However, unlike the situation in the U.S., where government support for the industry has come largely from the military and NASA, the European support programs are undertaken primarily by civilian agencies which have specific economic objectives. The position of the European IC industry remains rather weak relative to current competition between the U.S. and Japan. Thus its further role as an IC producer remains unclear.

In 1980, total European production of integrated circuits amounted to approximately $2 billion, while total consumption was $3.7 billion.

Philips

Philips holds the largest market share in semiconductors in Europe by a European firm.(18) It was the only major vacuum-tube producer in Europe which successfully made the transition to semiconductor devices. As mentioned earlier, it was one of the first European companies to acquire a Silicon Valley firm, Signetics. In IC's, Philips is attempting to obtain a greater coherence in its varied semiconductor operations. However, it is constrained by the scattered location of its facilities in many European countries whose governments exercise some degree of control over their operations.(19) For example, Philips has IC manufacturing operations in the Netherlands, the United Kingdom (Mullard), France (RTC), Germany (Valvo), and Switzerland (Faselec). The IC operating division is supported by a substantial corporate research activity, which has developed novel processes, devices, and manufacturing equipment.

Siemens

Siemens is the second-largest European producer of IC's, with production facilities in Germany and Austria. Corporate R&D laboratories form a significant part of the Siemens organization. The firm has followed a strategy which involves improving its own technological and production capabilities while simultaneously acquiring partial ownership in American firms such as AMD and several other high-technology companies.

Siemens is developing a strong capability in MOS technology.(26) The company's in-house expertise in power systems, telecommunication systems, and computers is reflected in its professionally oriented product mix.

ITT

ITT (International Telephone & Telegraph), although an American firm, produces most of its IC's in Europe, in particular West Germany, the U.K., and France. It is supported by an R&D facility in Germany. ITT is strong in the manufacturing of memory devices and is pursuing a market position in three growth areas: telecommunications, automotive applications, and EDP.

Smaller Manufacturers

Several smaller IC manufacturers are active in Europe. The following companies are listed by country:

- In the United Kingdom, Plessey, GEC-AEI, and Ferranti are mainly active as suppliers of some special circuits (Ferranti's gate-array type of technology, Plessey's work in GaAs and InP); and for their work for the British defense organization. A new company, INMOS, was started as a government initiative.
- In France, Thomson (Sescosem) is the largest IC producer. Several other French companies have been established, such as EFCIS, Eurotechnique, and Matra-Harris.
- In West Germany, AEG-Telefunken has a relatively small production of integrated circuits.

Several American companies have sizable industrial interests in Europe. We mentioned earlier ITT; but such firms as Motorola (France, the U.K., West Germany), Texas Instruments (France, the U.K., West Germany, Italy, Spain, Portugal) and National Semiconductor (the U.K., France) also have established European production centers.

THE JAPANESE INDUSTRY

In a relatively brief period the Japanese semiconductor industry has attained parity with the American in several product areas. The Japanese industry differs considerably from its American counterpart, operating in an entirely different framework. Government and industry cooperate in pursuing common goals. A number of Japanese firms cooperate jointly in areas of basic research.

Also, the financial environment of the Japanese semiconductor industry is quite different from that of the U.S.(20,31) One illustration of this is the cost of capital for a Japanese firm, which is only about 50 percent of the cost for an American firm. This is one of the consequences of the special relationship of Japanese banks with individual firms. This topic has been explored in depth in a study by the Chase Manhattan Bank.(21)

Japan's share of the world's semiconductor market has been increasing.(23,25,33) This is particularly true in markets for some of the most technologically sophisticated products. For example, the Japanese share of the American market for 16k RAM memories is over 40 percent. In the newer 64K RAM's it is estimated that Japanese firms had captured 70 percent of the world market in 1980.(32) There is evidence to assume that they will likewise be a major competitor in future memory generations.

These Japanese successes in the RAM business are embarrassing for the U.S. merchant industry, which used to develop its most advanced technologies for these products. The large volume of the market allowed them to recover such investments in a rather short time. Some prognoses therefore suggest that the U.S. IC industry will suffer heavily from this competition by Japan, even to the extent of a major shake-out and restructuring.(33)

As an explanation for this tremendous growth of the Japanese industry, some have pointed to the Japanese managerial focus on specific goals. Others think that the guidance and protection of the government has been a decisive factor. Nevertheless, most Japanese firms, although vertically integrated, concentrate on a narrow product line in semiconductors. In their marketing approach, also, they often target on specific American industries. In integrated circuits most of the Japanese production which is exported to the U.S. is in MOS logic and memories, intended for the data-processing market. The quality of Japanese IC's is generally considered to be a major competitive advantage. U.S. industry has devoted considerable attention to this issue, which has reportedly led to a closing of the gap. See Hinkelman's discussion for an American industry perspective.(6)

The economic framework in which these firms operate differs significantly from the American environment. J. Gresser has stated in an analysis(20) that "The key to understanding the Japanese semiconductor industry is its structure." Each major firm is a member of a corporate group, or Keiretsu. Firms within these groups cooperate in purchasing relationships and are interrelated in other areas such as finance or management. Japanese banks play an important role in these corporate groups.

The Japanese government, in particular represented by the Ministry of International Trade and Industry (MITI), through a series of laws and other means, has encouraged and coordinated the development of a domestic electronic/semiconductor industry. The most visible action has been the establishment of a common VLSI Laboratory in which industry and government participate.

The Japanese semiconductor industry is highly concentrated, being dominated by six firms: Nippon Electric Company, Hitachi, Toshiba, Matsushita, Mitsubishi, and Fujitsu. Sales of these six firms accounted for more than 80 percent of total semiconductor sales in Japan in 1979. In contrast, the nine largest American firms accounted for only 60 percent of semiconductor sales in the U.S. Unlike the American industry, Japanese firms are all vertically integrated and have been established in the industry for some time. This is reflected in their IC products, in which they developed strength in IC's for the booming entertainment industry. Only recently have they moved in to IC's for telecommunications and computers.(34) See fig. 6.4 for the 1979 production figures.

In recent years, several U.S. manufacturers (for instance, Texas Instrument, Intel, and Motorola) have set up production facilities in Japan. In addition to the advantage of producing IC's inside the duty barriers, these activities are seen as a good way to keep in close contact with Japanese technological developments.

On the other hand, Japanese companies are establishing large production facilities in the U.S., thus reducing their vulnerability to potential import restrictions. Examples of such companies are Hitachi, Fujitsu, and NEC.

Nippon Electric Company

NEC is the largest manufacturer of semiconductors in Japan. It is also a major supplier of telecommunications equipment and computers. The firm is particularly known for producing IC's for the telecommunications market. NEC has emphasized MOS devices and is presently strong in MOS Logic and MOS Memory which together comprise about two-thirds of the total production. It has acquired the American company Electronic Arrays. Recently NEC announced plans to set up a large production facility in the U.S.

Hitachi

Founded in 1910, Hitachi is presently the third-largest company in Japan. The firm is the nucleus of its own Keiretsu, the Hitachi group. Like other Japanese electronics firms, Hitachi is diversified and vertically integrated. The firm is the second-largest computer manufacturer in Japan, emphasizing mainframe computers and other nonconsumer electronics.

The firm was a participant in the VLSI project sponsored by MITI, in both computer development and semiconductor research. Hitachi is the second-largest semiconductor manufacturer in Japan. The firm is Japan's strongest bipolar producer.

Toshiba

Toshiba has traditionally been oriented toward the consumer market. The firm still generates over half of its semiconductor revenues in discrete products. However, recently Toshiba has become strong in MOS Logic. It has set up a joint venture with the new Silicon Valley startup LSI Logic to produce gate arrays in CMOS. The company is also expanding its foothold in the U.S. by acquiring a small semiconductor company.

Mitsubishi

Founded in 1910, the firm is part of the Mitsubishi Group. Mitsubishi is strong in MOS devices, which presently account for over 50 percent of production.

Matsushita

This firm is oriented toward consumer applications. More than 55 percent of its IC production is in linear devices. The firm has a large captive production.

Fujitsu

Japan's leading computer manufacturer is Fujitsu. Like the other major companies in the Japanese semiconductor industry, it was an important manufacturer of vacuum tubes. It is a member of the Dai Ichi Kangyo Bank group. The firm is the second-largest exporter of IC's to the U.S., next to NEC. Of the six major Japanese semiconductor firms, Fujitsu has the greatest percentage of its semiconductor sales, nearly nine-tenths, in IC's and is strong in MOS Memory. The company can be regarded as a strong competitor in the U.S. markets for the IC's and computers. It has set up various cooperative agreements with American companies (Amdahl, TRW) and European industries (Siemens) on computers and robotics.

104   U.S. MICROELECTRONICS INDUSTRY

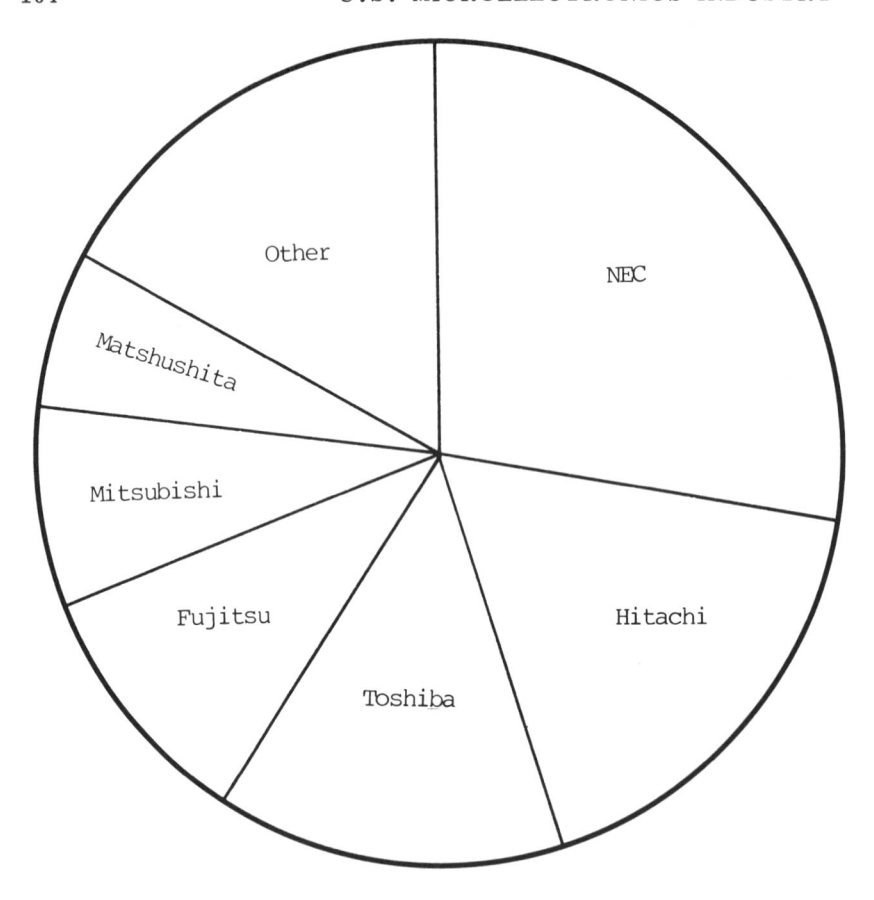

Fig. 6.4. <u>1979 IC production value in Japan.</u>

Source: Philips estimates.

REFERENCES

1. Dosi, Giovanni, <u>Technical Change and Survival: Europe's Semiconductor Industry</u> (University of Sussex: Sussex European Research Center, 1981).

2. United States Department of Commerce, <u>A Report on the U.S. Semiconductor Industry</u> (Washington, D.C.: U.S. Department of Commerce, 1979).

3. Federal Trade Commission, <u>The Semiconductor Industry: A Survey of Structure, Conduct and Performance</u> (Washington, D.C.: Federal Trade Commission, January 1977).

4. Wilson, Robert W., Ashton, Peter K., and Egan, Thomas, Innovation, Competition, and Government Policy in the Semiconductor Industry (Lexington, MA: D.C. Heath & Co., 1980).

5. Semiconductor Industry Association, 1979 Yearbook and Directory.

6. Semiconductor Industry Association, The International Microelectronic Challenge, The American Response by the Industry, the Universities and the Government (Cupertino, CA: S.I.A., 1981).

7. Gutmanis, I., Statement on Behalf of the Electronic Industries Association of Japan, U.S. Trade Commission, 332-107.

8. Noyce, R., Statement Before the Subcommittee on International Finance of the Committee on Banking, Housing and Urban Affairs, U.S. Senate, Jan. 15, 1980.

9. Jelinek, Marian, Institutionalizing Innovation: A Study of Organizational Learning Systems, (New York: Praeger, 1979).

10. Texas Instruments Annual Report, 1980; See also Texas Instruments, First Quarter and Stockholder's Meeting Report, 1980.

11. Uttal, B., "The Animals of Silicon Valley," Fortune, Jan. 12, 1981, p. 92.

12. Economist, "Innovative Intel," June 16, 1979, p. 94.

13. Harvard Business Review, "Creativity by the Numbers," May-June 1980, p. 122.

14. Intel Corporation, Annual Report, 1980.

15. Bloch, E., "Component Technology in a Vertically Integrated Company, A Technical and Management Perspective," Lecture Notes, Sept. 25, 1980 (IBM).

16. Eidson, J.C., "Fast Electron-Beam Lithography," IEEE Spectrum, July 1981, p. 24.

17. Fortune, "Europe's Wild Swing at the Silicon Giants," July 28, 1980, p. 78.

18. Business Week, "Philips, An Electronic Giant Rearms to Fight Japan," March 30, 1981, p. 86.

19. Center for Science and Technology Policy, "Status of French and German Electronics Industry," 1980, Unpublished.

20. Gresser, J., In a Statement to the Committee on Ways and Means, U.S. House of Representatives, Subcommittee on Trade, "High Technology and Japanese Industrial Policy: A Strategy for U.S. Policymakers" (Washington, D.C.: U.S. Government Printing Office, 1980).

21. Chase Financial Policy, "U.S. and Japanese Semiconductor Industries: A Financial Comparison," Report Prepared for the Semiconductor Industry Association, June 9, 1980.

22. Jones, K., "Europe Expected to Become a Net Exporter of IC's," Electronic Business, August 1981.

23. Bilinski, G., "The Japanese Chip Challenge," Fortune, March 23, 1981, p. 115.

24. Mackintosh, "Semiconductor Microlithography Equipment and Materials Outlook - 1985," Mackintosh Publications, U.K., 1981.

25. Pugel, T.A., et al., "Semiconductors and Computers: Emerging International Competitive Battleground," paper, Conference on "International Transfer of Resources: Strategic Company Responses In The Dynamic Asia Pacific Environment," Montreal, Quebec, October 1981.

26. "Chipmaker Siemens Plans to Ram Its Components," Economist, Nov. 28, 1981, p. 69.

27. "Competitive Factors Influencing World Trade in Integrated Circuits," U.S. International Trade Commission, USTIC 1013 (1979).

28. "Texas Instruments Shows U.S. Business How to Survive In the 1980's," Business Week, Sept. 18, 1978, p. 66.

29. "When Marketing Failed at Texas Instruments," Business Week, June 22, 1981, p. 91.

30. Le Boss, B., "Over A Million Devices Make Up 32-Bit CPU and Support Chips," Electronics, Feb. 10, 1981, p. 39.

31. "Japan's Strategy for the 80's: How Japan Will Finance Its Technology Strategy," Business Week, Dec. 14, 1981, p. 50.

32. Bylinski, G., "Japan's Ominous Chip Victory," Fortune, Dec. 14, 1981, p. 52.

33. "Confidential IC Report Triggers Public Debate," Electronic Business, September 1981, p. 120.

34. "Japan's Strategy for the 80's: Semiconductors," Business Week, Dec. 14, 1981, p. 53.

35. "Can Semiconductors Survive Big Business?" Business Week, Dec. 3, 1979, p. 66

36. Uttal, B., "Texas Instruments Regroups", Fortune, Aug. 9, 1982, p. 40.

7
Trends: A Look at the Future of the IC Industry

Having described the present status of technology, products, and industry, we have to consider some present trends that are likely to influence the future position of the U.S. microelectronics industry. The prospects of the IC industry are influenced by many factors; we will focus on several which are key. Obviously, scientific and technical factors are important determinants: will they allow the industry to continue originating new, economically attractive products that find wide applications in the world market? Evidently it is important to determine the major barriers in the future development of the technology.

But there are also nontechnical factors, which are equally critical. Among them is the important question of whether the industry is able to invest in the requisite R&D and capital goods to ensure its future expansion. In this rapidly growing business a company must have very good profitability to stay in the competitive race - if it does not, loss of market share occurs, usually progressively. It is not surprising that many companies have become absorbed in larger business structures that promise better stability and capital supply.

U.S. manufacturers encounter worldwide competition from countries whose governments have a substantial influence on their IC industry. Thus the profitability and future health of the U.S. industry are also determined by external factors such as the spectrum of public policies in the form of subsidies and other measures that exist abroad.

Finally, an interesting question is whether new ventures will continue to stimulate the growth of the U.S. industry as they did in the early seventies. This is an uncertain factor in view of the structural changes that have occurred throughout the industry. But several entrepreneurs have demonstrated their conviction that some specific industrial IC activities give fair promise of future expansion, even in a generally crowded competitive environment.

DEVELOPMENTS IN VLSI TECHNOLOGY

A prime question regarding the future development of IC technology is whether any fundamental or practical limits exist to a further increase in circuit complexity. Stated otherwise: can Moore's Law be expected to remain valid from a technological point of view when we are entering the VLSI era? More specifically, what can we anticipate in the next ten years in terms of circuit performance and price? A detailed answer to this question would require a thorough discussion of several technical aspects of IC's. This, obviously, falls outside the scope of this book. Other authors have outlined the present state of the technology and provided perspectives on this future.(1,2,20,21,25,27) We shall now discuss several of the more important issues.

In the next decade the resolution of lithographic techniques will be pushed below 0.5 micron. If currently 2-3 micron resolution can routinely be obtained, perhaps around 0.5 micron may be achieved on a similar basis in ten years. (Some particular parts of special IC's will be made much earlier with submicron resolution, as for GaAs microwave transistors.) Optical techniques, employing deep ultraviolet, may be extended below 1 micron, but it seems doubtful if much less is possible. Two methods promise a further reduction. X-ray lithography(3) has been developed at several laboratories. This technique of making patterns involves the use of a high-intensity X-ray source instead of a light source: the shorter wavelength of X-rays permits a better resolution to be obtained. This type of lithography required the development of several special techniques and materials, such as masks consisting of polyimide films with metal patterns and sophisticated photoresists.

Direct writing on the wafer with an electron beam still has an inherent better resolution, though scattering effects in the resist may appear to be a fundamental limit. The method is still too slow to be used in large-scale production. Work is in progress in several laboratories to increase the speed of the electron-beam machines and to enhance sensitivity of the resists. Recent examples are new machines developed by Hewlett-Packard(4) and IBM. Meanwhile, E-beam equipment is widely used for mask making. It may become a very attractive method for fast-turnaround pilot production.

In conclusion, it seems that several options may lead to the required lithographical techniques. However, the alignment of the subsequent patterns remains a critical factor.

Step-and-repeat optical methods are favorable in this respect. A direct-write E-beam technique may present certain advantages regarding alignment problems.

New etching methods are essential to further improve lithography. Classical wet etching techniques do not reproduce in silicon oxide the thin, crisp lines that one can make in the photoresist layer. Plasma etching techniques can be much better in this respect because they do not etch isotropically.

However, some major problems are expected to occur with the interconnection patterns. First, present design experience suggests that the length of the interconnections will increase disproportionately with increasing complexity of the circuit. These longer connection paths are narrower and thus represent a higher resistance in the circuit. Because signals travel longer, there may be a loss of synchronization in the circuit. Such effects on circuit behavior must be carefully evaluated.

Additionally, electromigration (displacement of metal atoms under influence of electric fields) must be tightly controlled. Work is being undertaken to develop new types of interconnect, and several materials show good promise.

A further problem arises when at greater complexity the number of pins available in the package is insufficient to accommodate all needed input and output connections. In this respect it should be noted that in most merchant IC designs, little attention has been devoted to packaging technologies. An example of a novel approach has been given by IBM, which developed a ceramic substrate with multilayer connection on which the IC is mounted. One may expect that such technologies will find a wider application.

Heat dissipation in the substrate may become a problem when the circuit density is increased further. The use of CMOS technology may provide solutions in many cases where circuitry is not continuously switched, so that mainly steady-state dissipation occurs. Also, better mounting and cooling techniques may be devised, which may largely eliminate the dissipation problem.

The size of the chip has to be increased further to keep up with the predictions of Moore's law. This expansion will require an increased mastery of the silicon material used for the wafers. The electrical, chemical, structural, and crystallographic properties of the wafer have to be well understood and carefully controlled. One should keep in mind that silicon growth is still largely an empirical process, and a better understanding of the theory may lead to substantial improvements in the control of the material properties.

From the viewpoint of solid-state physics, the envisaged reduction in transistor size is not expected to lead to major difficulties. It must be emphasized, however, that computer models used in transistor design may still prove more or less inadequate at the small dimensions considered. Consequently, further studies will be needed to verify their validity. Unexpected physical phenomena, not included in any model, may always turn up. For instance, the alpha-particle problem in the 16k RAM appeared to be a serious obstacle. A solution was found, however, and we may have confidence in the ingenuity of the IC developers when other problems arise.

A major problem is presented by the need to test VLSI circuits. Conceptually, this must be solved mainly in the design phase, but the physical methods of testing the very small circuits need a high degree of sophistication.

Summarizing, from a technological point of view there seems to be no immediate reason to assume that Moore's law will break down. However, it is likely that progress will be somewhat slower since techniques become progressively more difficult with increasing complexity.(5) This can be illustrated by the differences between 64-kb and 256-kb RAM technology.(32) Much work in many areas is still ahead, but no real barriers are in sight. The speed of increase in complexity may be seriously slowed, of course, if problems emerge that are more severe than expected. In any case it now seems that the design techniques will present a more difficult problem then the technology.

It is not expected that this situation will seriously affect the industry's growth potential. A certain lag in the application of IC's has occurred, and a period of slightly slower technical advances might even seem desirable in some application areas. However, technical improvements remain the name of the game: no major company can permit its technology to slip with respect to that of the competition.

VLSI DESIGN

A major obstacle in the road toward very-large scale integration is the state of current design technology.(22,31) In recent years design times and costs have been rapidly increasing as circuits have increased in complexity. If this trend is to continue, only a few generally applicable circuits and some specific high-volume chips will be made; for smaller series, development costs are prohibitive.

However, a need exists for low-volume customized circuits for use in specific systems. One can argue that this need will grow when IC complexity increases. Many captive activities in systems houses are an indication of their interest in custom IC's. Though for them the cost per IC is a somewhat less critical factor than it is for merchant houses, an economic and convenient VLSI design technique is highly desirable.

It can also be argued that increasing circuit complexity will change the cost difference between the application of a general-purpose IC and a special-purpose design. General-purpose IC's will have a relatively large size if they are designed to cover many applications, which adds to the manufacturing costs. Moreover, indirect costs of their usage such as software are also known to increase strongly with growing complexity. Thus the development of a custom IC may become a more attractive alternative from the point of view of total costs: higher design costs may be offset by lower manufacturing and application costs. Clearly, the future availability of advanced VLSI design techniques and the necessary fast, easy-to-use computer-based systems will strongly influence this trade-off in total costs incurred.

It is not yet clear which option is best in VLSI design.(6,7) Some examples of continuing research in this field may serve to illustrate the complexity of the problem.

Present IC Designs

The present IC's which exhibit the highest degree of integration are memories and microprocessors. The advantage of memory designs is that they consist of many repetitions of the same cell structure. Though by no means negligible,(5) this type of circuit is not representative of the problems of VLSI design.

The microprocessor is probably today's most sophisticated IC design. Its high design costs are amortized by large-production series, thanks to a wide application range. However, the deployment of a microprocessor in specific applications requires substantial software expenditures borne by the user. The problem is how to optimize hardware and software expenses on a systems level.

Gate arrays have become very popular of late in the computer industry. It seems likely that their "quick and sloppy" design method will remain attractive for several years to come. Current CAD methods for gate arrays are inadequate to handle large numbers of gates, but they may improve. Ultimately, improved design methods should yield large-scale customized IC's with a reasonably optimized circuit design at acceptable costs.

VLSI Design

A VLSI-design system should be compatible with computerized systems-design techniques. For the time being, however, it is assumed only that systems design has led to the specification of a function which is to be incorporated in one or more VLSI chips. A VLSI-design process can be considered at various levels of abstraction. At the highest level, one decides how VLSI circuitry can be applied to perform a specific function. For instance, should a given picture-processing task be carried out by a microprocessorlike circuit, by a dedicated computer chip, or with a highly parallel processor?

When the decision has been made, the architecture of the IC (or IC's) has to be considered. Questions like how the circuit will be separated functionally or which data-communication structure is required will have to be decided. These choices strongly depend on the design strategy that is adopted. For instance, one may prefer certain classes of solutions in view of the availability of computer tools for design and simulation. But one may alternatively choose a manual method because it results in an optimal use of silicon area.

Then the actual circuit design must be carried out. Starting with logic equations describing the circuit, networks of "and"-"or" gates can be designed. The logic circuits must be chosen and special circuits like output drivers must be conceived. The layout phase includes such activities as positioning transistors and interconnections on the chip according to certain design rules and checking the layout for correctness with respect to specifications. Finally, mask tapes are generated and the IC's can be produced.

This "layered" structure does not necessarily imply a strict top-down approach, in which the design is begun with quite general specifications which are successively detailed in later stages. The possibilities offered by a silicon technology finally used should have great influence on the choices made at higher abstraction levels, actually even at the systems-design level. The importance of having a good overview of all design stages has been stressed in the methodology proposed by Mead and Conway.(8)

In a hierarchical approach the same circuit is treated at several levels of abstraction. This process necessitates:

- Descriptive languages at various levels (for instance using Boolean algebra for the logic description), along with clear interface definitions (e.g., in the Caltech Intermediate Format language).
- Front-end tools, those which enable the user to work with the various representations (e.g., interactive computer graphics for stick diagrams).
- Tools to be used in various levels such as transistor models, simulation programs, timing checkers, or verification techniques.
- A data base in which various representations can be stored and kept consistent.

The tools that are available now are mainly used at the lower levels of abstraction. Efforts to develop tools for the higher levels are described in the next section.

A major problem in designing VLSI is testability. This raises the issue of how one can establish whether the performance of a circuit is in agreement with its initial specification.(7) A general approach is not available, but apparently many design errors can be eliminated by applying a clear and structured design discipline to obtain "correctness by construction." Research efforts here include the use of the same program for the simulation of the design and the test of the chip.

VLSI-Design Techniques

The major task in the VLSI-design process is to master complexity. This will entail a restriction of the possible paths which in principle can be followed in the design process.

A very rigorous approach along those lines has been proposed by Mead and Conway.(8,23) They use an n-MOS technology with simple, rather relaxed design rules. Also, a two-phase timing is chosen, combined with a systematic "data path" structure consisting of a series of register-logic-register stages. Often a programmable logic array (PLA), a regular structure consisting of arrays of "and" gates and "or" gates, can be applied to perform the logic function. In this way, highly regular structures can be made with a minimum of interconnection lines. This seems to be a favorable strategy, since in many VLSI designs the area covered by the interconnections is expected to exceed that occupied by the transistors.

An advantage of this approach is that the logic function can be implemented in a PLA with the help of computer programs. Software may also be used to generate the actual layout of the circuit, with given standard cell structures. The desired function is described in a high-level language, and a "silicon compiler" takes care of the translation into a layout design.

Such compilers may be designed for other functional blocks as well: for instance, a multiplication circuit might be computer-generated from some initial specifications such as word lengths to be used. At the California Institute of Technology a certain kind of microprocessor structure has been computer-designed from a set of initial specifications.

The abstraction level of such software tools is high, but the generality is rather low: only one particular class of circuits can be produced.

The criticism leveled at this approach, which has been primarily followed at universities, is twofold:

• These methods use very unsophisticated transistor models and simple design rules and are wasteful in terms of

circuitry. They result in large, unoptimized designs which are not attractive for merchant houses, whose competitive position often depends on as optimized and small a circuit as possible.
- The methods have been applied to "toys," relatively small-scale designs. It is by no means certain that they will prove to be useful for VLSI.

Though currently rather controversial, the underlying ideas seem to be basically sound and will probably find many applications in the more systems-oriented industry.

An alternative approach to the design problem consists in defining a set of building blocks, which are designed once, then stored in a computer and assembled by a designer into bigger units. Thus, if it were possible to define a reasonably complete set of building blocks and interfaces, a company could efficiently integrate and apply its design experience.

A further abstraction of the design process might involve the introduction of software concepts. Structured design practices, for instance, have proven their value. Similar techniques might be useful in designing VLSI. This application, however, will require a quite different approach of the design process.(9)

Finally, we should mention ideas to use artificial intelligence concepts for VLSI design. Analogous to "expert systems" built, for instance, in the medical field,(10) such programs could attempt to reason and act like human designers.

CAD and Software: Analysis

Currently design methodology seems to be the major obstacle on the road to VLSI. Many CAD packages exist, but most will not be suited for VLSI. Furthermore, they are unrelated and do not constitute a coherent "CAD system." It is not very clear which methods will emerge as new effective tools. Compared with technology development, relatively little effort is spent on these problems.

Design methodology can be influenced by new technological developments. For instance, general availability of a multilevel interconnection technology will considerably ease the design process. Obviously, new technologies will also necessitate new design techniques.

Two observations can be made on the similarities between software development and VLSI design. First, high-level languages have increased the efficiency of software design, though they usually offer the programmer only a subset of all possible combinations of instructions. Admittedly, the final codes may use more memory space than a program written in machine code, but with the current machinery this limitation poses no obstacle. Additionally, errors are found only in the program (written in "understandable" high-level language) and no longer in the machine code, resulting in the elimination of many trivial coding errors. Similarly, high-level descriptions of VLSI combined with the use of compilers may lead to highly increased design productivity. This advantage may be much more important than the restriction in design possibilities imposed by the language or the fact that the final design is not optimal in its usage of silicon. Design errors here will not occur on the gate level any more but only in the high-level language description, improving the correctness of the design.

The second observation concerns the introduction of the Mead-Conway methodology in several universities (see chapter 8). This discipline has enabled computer scientists to design their own IC's. Chips were made from these designs and used in the intended applications. There is a resemblance here to the situation in physics of about twenty years ago, when physicists began to make their own software programs without the help of an intermediate programmer. This new practice has substantially altered the way physics research is conducted. One may foresee a similar influence of Mead-Conwaylike methods on computer science, where at present rather little feedback is available on the practicability of certain architectural ideas.

SEVERAL CHALLENGES

The trend toward smaller dimensions will also influence VLSI architecture. With smaller transistor size, higher operating frequencies, and larger chips, new architectures become practical. However, new limitations will also become evident. For instance, synchronization may not be maintained over the full surface of the IC, and novel designs will be needed to cope with this problem. Another example will be the need for fault-tolerant structures, made possible by the great complexity available but also made necessary by the same complexity. In this way, a greater interaction between design, technology, and physics will arise, at any rate temporarily counteracting Mead-Conway's trend toward separation of design and technology.

As a consequence, much R&D will be necessary in these areas. New concepts are needed, which academic research may provide. An increased interaction between systems aspects and IC design will be necessary, possibly giving new stimuli to the vertical integration trends of the industry.

STRUCTURAL DEVELOPMENT OF THE U.S. INDUSTRY

The major development of the U.S. integrated-circuits industry has been an increased vertical integration in the last few years, with only a few new entrants. Financial as well as technical reasons explain this phenomenon and will strongly influence the future development of the industry.

Capital Investment

The capital investment needed for an economically viable production facility has been rising rapidly in the last decade.(17) Ten years ago a production unit of 10,000 3-inch wafers per month was necessary for a sales volume of $25 million. Currently, a 20,000, four-inch wafers/month facility for a more than $80-million sales volume is seen as a minimum requirement. This is an expensive entry ticket for a new company wishing to acquire a wafer facility.

The necessity for this scale in growth follows from the fact that the market price per circuit (in constant dollars) drops annually by about 9 percent. Combined with a 9 percent average inflation rate, industry is forced to improve productivity by at least 18 percent per year. This has been done successfully, on the one hand by increasing the number of circuits per square inch and on the other hand by producing larger wafer diameters and more efficient equipment. This modus operandi, however, has led to a steep increase in capital-investment needs: in 1970-75, approximately 40 cents was needed to obtain a dollar in new sales; in 1975-80 this increased to 55 cents, while currently as much as 70 cents is needed for a future dollar of new sales.

Another factor is related to the strong growth rate of the industry (say 30 percent per year), which requires large investments in new production equipment. A decision now to invest in a factory will lead to production in three years. On the average, in the current year investments must be made for enlarging production facilities over two years. At that time expected annual sales growth will amount to 39 percent of the current sales. As we have seen, 70 percent of this sales growth must be invested now in capital goods. Therefore, approximately 27 percent of current sales is necessary for capital investment for future expansion.

This scale of expenditure generally cannot be entirely financed internally, since depreciation is usually 8 to 10 percent of sales, and profits after taxes of 5 to 12 percent add up to only 15-20 percent of sales available for investment. It is unusual among new IC firms to pay dividends to stockholders. New stock can be issued or bank loans may be obtained, but the maximum debt-equity ratio allowed by banks is usually limited. These financial burdens have put the independent IC makers in a vulnerable position. Success is essential for survival, but even relatively minor commercial errors or a somewhat prolonged economic recession can endanger the continuity of the companies. Thus it does not come as a surprise that many companies have been acquired by larger entities (e.g., Fairchild by Schlumberger, Signetics by Philips, Mostek by United Technologies, and Intersil by GE) and have thus obtained the necessary financial backbone and stability to ensure future operations.

The vulnerability to economic recessions can be illustrated by what occurred in 1974. Investments were rigorously cut and personnel laid off, resulting in undercapacity when business picked up again. This contraction is generally viewed as the reason why Japanese producers, who had not curtailed their operations so strongly, could capture a large market share in the years following the recession. During 1980-81, most U.S. companies were careful not to repeat their 1974 mistakes, but this course stressed their short-term profits.

Systems Aspects

From a technical point of view, integrated circuits are becoming so complex that many require appreciable systems know-how in the design phase. A striking example in this respect is the microprocessor. Its performance has increased so strongly that it is now competing in performance with minicomputers and even larger mainframes. Developing such microprocessors has required a systems approach to hardware and software that was previously unknown in the industry. Another example is the design of telephone circuits like SLIC's and Codec's, which require a detailed knowledge of the operating conditions of telephone systems in many countries. The IC industry's need for specialized knowledge will increase with the growth in application of microelectronics technology.

Systems houses, on the other hand, see gradually moving more and more of their system into the chip. For various reasons, such as proprietary information leakage, keeping added value in-house, and certainty of logistics, they may wish to concentrate chip design and manufacturing largely in their own companies.

Consequently, the merchant IC industry will have a tendency to broaden its technology base toward the areas where its products are applied. This may be effected via cooperative agreements with systems houses or takeovers, for instance of software firms. A forward integration trend is obvious. On the other hand, many systems houses are hence interested in backward integration. They achieve this via take-over of IC manufacturers or by setting up captive activities. In conclusion, a change in the industry from independent, monolithic companies into a more integrated pattern of larger units is evident.

Many companies have realized that there is usually more added value in systems manufacturing than in IC production (and also that it reduces the vulnerability to economic changes). The result is greater diversification. Texas Instruments is a prime example, with a management system that puts heavy emphasis on innovation and diversification. Intel has been strongly promoting its systems activities, based on microprocessors and the software and development tools that go along with them. Another example is National Semiconductor, which has been active in watches and calculators in the past. It has now diversified into "point-of-sales terminals" for stores and into computers.

In view of these considerations we may expect further integration of the IC industry in the forward direction. (Backward integration, i.e., in manufacturing equipment or base materials, has been actually diminishing.) Since merchant houses need large amounts of money to finance their future operations, they will have little money available to integrate forward any further. But large systems companies will presumably take over several of the remaining independent merchant houses, particularly when the latter experience structural or incidental problems which threaten their growth.

Another trend will be the development of business structures needed for relatively small-scale custom VLSI. Separate houses for design, maskmaking, and process technology will be established, providing a wide range of services for the (smaller) systems companies. In addition to new ventures, several established IC houses like Intel, National Semiconductor, and Signetics have engaged in "silicon foundry" operations. They will make part of their manufacturing technology available to process chips designed by other firms.(28)

The foregoing discussion has dealt mainly with the relation between the IC industry and the "conventional" electronics business. However, microelectronics will increasingly be used in presently nonelectronic industries. On the one hand, considerable business opportunities seem to be available for the IC and equipment industry. However, novel ways of dealing with these nonelectronic industries have to be developed. New business structures adapted to these particular markets may well shape up.

NEW VENTURES AND INNOVATION

After an absence of new ventures for about five years, nearly a dozen new IC-fabricating companies have been established since 1980. This development can be partly explained by financial considerations. In the U.S., renewed availability of venture capital has occurred since 1978.(18) In that year, the capital-gains tax was reduced to a maximum rate of 28 percent, from an earlier 49 percent. In 1981 it was reduced again to a rate of 20 percent. Another factor is strong public interest in high-technology stocks. Young companies can go public at high price-earnings ratios, giving venture capital companies the opportunity to earn meaningful returns on their equity investments. Moreover, high inflation rates make many competing investment forms unattractive.

New opportunities have been perceived by entrepreneurs, notably in catering to the custom-IC market, which has been somewhat neglected by existing companies. Thus nearly all new ventures intend to deliver some form of custom IC's (either gate arrays or specially designed circuits) or services (such as "silicon foundry" operations(28)). Much will depend on how they will interface with the customer. Availability of software design tools (such as needed for gate-array design) or a fast, standardized design and manufacturing service (e.g., based on the Mead-Conway design methodology) will be a determining factor for success.

Only a few new ventures seek direct competition with existing companies, either on the basis of a mastery of a technology (e.g., MOS) or on a superior design of a specific device (such as a novel EEPROM). As noted in the foregoing section, the necessary capital investment is high - much greater than the $5 to $10 million reportedly necessary to start up a company in the former category. Moreover, the advantage has to be capitalized quickly, since existing firms move their technology and designs steadily forward and are bound to outdistance a newcomer who cannot keep ahead.

A new idea has been made public: the deliberate establishment of a company that would develop very advanced technology. IC makers would commission such developments to this company on a normal commercial basis.(30)

Indeed, newcomers may have a very hard time. It is true that more room exists in the custom area, but it is unclear whether the market will be large enough to keep all those companies viable. It seems a fair guess that particularly the new custom-IC manufacturers will become attractive acquisitions for systems companies seeking to improve their capabilities for proprietary design.

In the past, innovation in the industry has been strongly connected with new ventures.(16,29) It is customary to orient a firm's interest in product innovation when it is young and growing, whereas in larger, mature companies the tendency is to emphasize process innovations. Thus it is interesting to observe whether the innovation rate in the industry is slowing down now that the major participants are developing into large, world-wide operating firms.

In effect, much of the innovation is already directed toward the slow, steady improvement of processes to keep costs at a competitive level. On the other hand, the rapidly expanding market in several areas for the use of integrated circuits suggests that excellent business opportunities will remain for those companies which come with innovative designs. In such an atmosphere, most companies will be compelled to adopt an innovative behavior, though perhaps not to the extent that the company's future is placed in jeopardy by concentration on a novel circuit.

A major question is whether entirely new techniques will make the current IC technology obsolete. For instance, integrated optics might replace digital IC's as the basis for computers. Such a development cannot be ruled out, of course. However, at present vast areas of application would still remain wide open for the current technology and its future derivatives. Consequently, much efforts will be spent on "conventional" technologies, perhaps leaving little motivation for active pursuit of basic replacement technologies. A few areas exist, of course, where the current technology is pushed close to its limits (for instance, in superfast computing). These areas should be monitored for the emergence of new technologies.

GOVERNMENT SUPPORT STRATEGIES

Many nations began assisting their integrated-circuits industry when its contribution to economic growth and export competitiveness became apparent. Though concerns for national security were partly involved, in the majority of cases support was given in the framework of an industrial policy. The philosophies and methods of various governments differ widely. Some typical examples of subsidies and loans to IC business abroad are summarized in table 7.1.

France

In France, the Ministry of Industry, the General Directorate for Telecommunications (DGT) and the General Delegation for Weaponry (DGA) have been reinforcing the high-technology base of the nation's industry for many years. In particular,

Table 7.1. Commercial Government Subsidies and Loans for Semiconductors

PROPOSED & ACTUAL

COUNTRY	MAIN RECIPIENT	STATE FUNDS ($M)
GERMANY	VLSI DEVELOPMENT	100 (30-40 YEARLY)
ITALY	SGS-ATES, et al.	135
FRANCE	ST. GOBAIN PONT A MOUSSON MINISTRY OF INDUSTRY TO THOMPSON CSF-SSC THOMPSON CSF + CEA (SESCOSEM/EFCIS) RADIOTECHNIQUE COMPELEC MATRA	50 120 TO 200 25 38
UK	UK TOTAL: $330M NEB AVAILABLE FUNDS LIMIT: $6B NEB TO INSAC (SOFTWARE CONSORTIUM) NEB TO INMOS LTD. NEB TO PLESSEY (LOAN) DOI TO MISP (MICROELECTRONIC INDUSTRY SUPPORT PROGRAMME) DOI TO MAP (μP APPLICATIONS PROJECT) E-BEAM FAB TECHNIQUES NCC - AWARENESS PROGRAM (SOFTWARE TRNG)	 40 90 40 140 110 1.8 20
JAPAN	VLSI SUBSIDY (LOAN)	250
KOREA	TOTAL PROJECTED LOAN & SUBSIDIES GOLD STAR KIET - WORLD BANK LOAN	600 20 29
	TOTAL GOVERNMENT EXPENDITURES WORLDWIDE FOR PROMOTING SEMICONDUCTORS	2.0B

Source: "Impact of Semiconductor Nationalism on American Trade, Technology and Defense," Terry Wong, Rockwell International, April 1979, as cited in Hearing before the Subcommittee on International Finance of the Committee on Banking, Housing and Urban Affairs, United States Senate, "Trade and Technology in the Electronics Industry," Jan. 15, 1980, p. 132.

the development and industrialization of computers and telecommunications have received substantial financial support.

Some five years ago greater appreciation arose for the interdependence of systems and integrated circuits. In 1977 this led to the establishment of the Plan Circuits Intégrés (or Plan Composants) which is aimed at strengthening the domestic IC industry. The sum of $120 million is involved, the major portion of which is directed toward obtaining a domestic MOS capability in France. This plan sets forth R&D targets for both government and private laboratories, production targets for large-volume circuits, and production targets for custom-built integrated circuits for French users. Five firms are involved in the program:

- Thomson's semiconductor division, Sescosem, technologically supported by Motorola (mainly bipolar technology).
- Efcis, a joint venture of Thomson and CEA, also with technological ties to Motorola (MOS).
- RTC, a Philips subsidiary (fast bipolar technology).
- Eurotechnique, a new joint venture of Saint-Gobain and National Semiconductor (MOS).
- A new joint venture by Matra and the U.S. company Harris (MOS).

In these joint ventures, American companies are restricted to a minority position (49 percent in both cases), as required by the French government. However, not all American high-technology companies are willing to exchange their technology for access to the French professional market.(11)

A continuing R&D program, currently funded at $14 million per annum, covers a broad range of IC-like technologies and applications. The work is mainly done in several government institutes such as LETI, operated by the French Atomic Energy Commission, CNET, a PTT laboratory, and in industrial laboratories.

Research efforts in the universities, for instance Toulouse, Montpellier, and Paris VI, are often considered to have remote ties with industrial development. Some regional interuniversity technology centers are being established in Toulouse, Grenoble, and Paris.

The future development of the French IC industry is currently rather uncertain. The present government has nationalized several of the companies involved in the IC program. In such a scheme, the different programs might be better coordinated and more heavily funded. This would be in line with the government's R&D policy, which calls for heavily increased spending over the next few years. Indeed, the 1982 budget and published plans indicate a further strengthening of the support of the microelectronics industry.

West Germany

The government's support for microelectronics is channeled mainly through the Federal Ministry for Research and Technology (BMFT). It conducted a program for electronic components development in the years 1974-78, an effort which is being continued. The general aim of the program is to establish the technology base needed to bolster German industry in preparation for its future international competition and to acquire the capabilities for the development of the requisite technical infrastructure of German society.

Programs are concentrated on R&D in products and processes which hold clear industrial promise. In addition, the programs aim at strengthening the necessary industrial structures and improving the industrial innovation process.

Major areas of emphasis relate to VLSI design, such as lithography and technology, and optoelectronics. Both underlying basic research and applications are included in the program, which amounts to roughly $44 million annually. A notable area of research is lithography with synchrotron radiation in connection with the new BESSY facility in West Berlin.

Contracts are awarded to universities, R&D institutions, and industrial laboratories on the basis of their proposals. BMFT influences the actual R&D programs to obtain the necessary coherence between different efforts and to promote further industrialization of the results.

Some major programs in VLSI are conducted by the Universities of Aachen and Dortmund, the Fraunhofer Institute in Munich, the Max Planck Institute in Stuttgart, and also by major industrial laboratories.

United Kingdom

In the United Kingdom a rather unorthodox approach has been undertaken to achieve technical competence in microelectronics technology. In 1978 the National Enterprise Board, a government agency, promised a $120-million investment in INMOS.(12) This company, founded by two entrepreneurs, aims at producing and marketing various types of MOS memory. Its design and technology development takes place in the U.S. (Colorado), whereas INMOS intends to have production facilities in South Wales. INMOS is currently marketing its first products and the degree of success is still highly uncertain.(24)

The Department of Industry, in the interim, supports improvement of the British industrial facilities via its Microelectronics Industrial Support Programme, which is funded with approximately $93 million annually. In addition, the department has established a Microprocessor Application Project, funded with a similar amount. Under this program British industry can obtain subsidies for the acquisition of the knowledge necessary for the use of microprocessors in their manufactured equipment. The aim is to improve the international competitive position of products from British industry.

Japan

The Japanese government has had a long history of supporting the development of its electronics industry.(13,14) An overview of some programs is given in table 7.2. Initially programs were designed to close the gap with other industrialized countries. In addition to R&D subsidies, temporary tax barriers against imports were created, favorable loans and tax measures for capital investment were granted, and export subsidies were given. Later on, the Japanese government chose electronics and especially information technology as one of the industrial pillars of the future society. This implies that Japan must assume a leading world position in such technologies. The government actively coordinates, usually via MITI, Japanese efforts to achieve such a position.(26)

As an example, a "Machinery Information Law," covering a period from 1978 to 1985, contains measures to promote a Japanese information industry and a high-level manufacturing technology. It contains programs to advance research and development in a large number of disciplines, one to support their industrialization, and one aimed at providing the necessary fabrication technology. The mechanisms chosen are loans for capital investment, tax relief for special investments, and credit facilities for modernization of the factories.

Some special subsidies were established to develop a Japanese computer industry. In this framework the development of an IC industry was strongly pushed. The establishment of the Japanese VLSI program has received worldwide attention. MITI coordinated the establishment of a joint VLSI laboratory in 1976. Fujitsu, Hitachi, Mitsubishi, NEC, and Toshiba took part in this $140-million, four-year effort. The goal of the project was to establish a Japanese position in microprocessors, design techniques, processes, testing, performance evaluation, and devices. The impact of this program on the Japanese MOS technology has been appreciable.

The VLSI program forms part of a much broader program for the development of computers and telecommunications equipment. Joint government-industry laboratories were established under this program. One group, composed of Fujitsu, Hitachi, and Mitsubishi, operated a jointly owned company established to develop IBM-compatible computers. A

Table 7.2. Some Japanese Government
Semiconductor Programs

1966-71	Super-high performance computer system	$ 28 M
1971-78	Pattern information processing system	$ 97 M
1973	Process technologies program	$ 20 M

        Silicon gate   Hitachi
        CMOS   Toshiba
        Bipolar digital   Fujitsu
        NMOS   NEC
        Industrial linear NEC

1975-77	NTT Memory Development	$ 67 M
1976-80	VLSI Project	$310 M*

VLSI Technology Research Association
    Cooperative Laboratories
    Computer Development Laboratories
        (Fujitsu, Hitachi, Mitsubishi)
    NEC-Toshiba Information Systems Laboratories
        (NEC, Toshiba)

---

*Total budget; split 40 percent government, 60 percent industry.

Source: W. M. Bullis, "Government Programs on Advanced Technology and Manufacturing Techniques: Comments on U.S.A., Japan and Europe," National Bureau of Standards, June 1978.

second group consisting of NEC and Toshiba was involved in developing computers not compatible with IBM. Present government efforts are oriented toward research in the use of Josephson junctions and the use of gallium arsenide in IC's.
    Currently, an effort on a "fifth-generation computer" is being started, aimed at nonconventional computer structures.(15) For this program, MITI seeks a form of cooperation with U.S. and European universities and manufacturers in view of the very advanced character of the R&D involved.
    It should be noted that such Japanese programs form part of a comprehensive effort to master information technology. Programs exist in software development, pattern recognition, intelligent manufacturing, etc. Special govern-

ment laboratories (Tsukuba Science City) have been set up to carry out the necessary investigations.

Other Countries

Several other European governments have embarked upon modest efforts to support their domestic microelectronics industries. Also, programs oriented toward applications of the technology in industry and society have been established. In Sweden, for example, a government support program currently provides about $3-4 million annually for the domestic components industry. In 1981 the Dutch government initiated a program for the promotion and application of microelectronics.
    The Commission of the European Economic Community has been developing proposals for support of industrial R&D directed toward VLSI. Design technology and manufacturing equipment will be the areas to be supported. Subsidies amounting up to 50 percent of the R&D costs incurred by the industry totalling approximately $140 million are planned to become available in the period 1981-84. In this way the Commission hopes that the European industry will get a better position in the struggle with the U.S. and Japan. In 1981 the EEC Commission took the initiatives for discussions with governments and industry on a Long Lead Time industrial R&D program, in which microelectronics will have a central role (the ESPRIT program).

REFERENCES

1. Tarui, Y., "Basic Technology for VLSI, Part II," IEEE Trans. on Electron Devices ED 27 (1980) 1321.

2. Keyes, R.W., "The Evolution of Digital Electronics Towards VLSI," IEEE Journal of Solid State Circuits SC 14 (1979) 193.

3. Lepselter, M.P., "X-ray Lithography Breaks the Submicron Barrier," IEEE Spectrum, May 1981, p. 26.

4. Eidson, J.C., "Fast Electron-Beam Lithography," IEEE Spectrum, July 1981, p. 24.

5. Berhard, R., "The 64-kb RAM Teaches a VLSI Lesson," IEEE Spectrum, June 1981, p. 38.

6. Robinson, A.L., "Are VLSI Microcircuits Hard to Design?" Science 209 (1980), 258.

7. Capece, R.P., "Tackling the Very-Large Scale Problems of VLSI," Electronics, November 23, 1978, p. 111.

8. Mead, C., and Conway, L., Introduction to VLSI Systems, (Reading, MA: Addison-Wesley), 1980.

9. Allen, J., and Penfield, P., "VLSI Design Automation Activities at MIT," IEEE Trans. on Circuits and Systems, 28 (1981) 645.

10. U.S. Department of Health, Education and Welfare, National Institutes of Health, "The Seeds of Artificial Intelligence: Sumex-Aim," Bethesda, MD, 1980.

11. Business Week, "Why Intel Shies Away from a French Deal," March 2, 1981, p. 28.

12. Science, "INMOS Enters the 64K RAM Race," May 8, 1981, p. 642.

13. Japan Electronic Industry Development Association, "Future of the Japanese Electronic Industry," 1980.

14. Gresser, J., In Statement for the Committee on Ways and Means, U.S. House of Representatives, Subcommittee on Trade, "High Technology and Japanese Industrial Policy: A Strategy for U.S. Policy Makers" (Washington, D.C., 1980).

15. Lehner, U.C., "Japan Starting 10-Year Effort to Break Exotic Computer," Wall Street Journal, Sept. 25, 1981, p. B1. See also Manuel, T., "West Wary of Japan's Computer Plan," Electronics, Dec. 15, 1981.

16. "Can Semiconductors Survive Big Business?," Business Week, Dec. 3, 1979, p. 66.

17. Finan, W., "The Semiconductor Industry's Record on Productivity," Technecon Analytic Research.

18. Pollack, A., "Few Places for Venture Capital," New York Times, June 17, 1981.

19. Abernathy, W.J., and Utterback, J.M., "Innovation and the Evolving Structure of the Firm," Harvard Business Review, HBS 75-78, 1975.

20. Sze, S.M., "Semiconductor Device Development in the 1970's and 1980's - A Perspective," Proceedings IEEE 69 (1981), 1121.

21. Lohstroh, J., "Devices and Circuits For Bipolar (V)LSI," Proceedings IEEE 69 (1981), 812.

22. Newton, A.R., "Computer-Aided Design of VLSI Circuits," Proceedings IEEE 69 (1981), 1189.

23. Marshall, M., et al., "The 1981 Achievement Award," Electronics, Oct. 20, 1981, p. 102.

24. "Nexos, Inmos, Off with Their Heads," Economist, Nov. 28, 1981, p. 76.

25. Lyman, J., "Scaling the Barriers to VLSI's Fine Lines," Electronics, June 19, 1980, p. 115.

26. "Japan's Strategy for the 80's: Japan Inc. Goes International with High Technology," Business Week, Dec. 14, 1981, p. 40.

27. Cooper, J.A., "Limitations on the Performance of Field-Effect Devices For Logic Applications," Proceedings IEEE 69 (1981), 226.

28. Arnold, W.F., "Silicon Foundries: Custom Work, Mass Price," Electronics Business, November 1981.

29. "Innovation in Small and Medium Firms," Background Reports, OECD, Paris, 1982.

30. "A For-Profit Lab to Help Chip Makers Compete," Business Week, April 5, 1982, p. 35.

31. Trimberger, S., "Automating Chip Layout," IEEE Spectrum, June 1982, p. 38.

32. Bernhard, R., "Rethinking the 256-kb RAM," IEEE Spectrum, May 1982, p. 46.

8
The Technology Base of the U.S. Integrated Circuit Industry

A technologically sophisticated, high-growth industry such as the U.S. integrated-circuit business depends heavily on the quality and quantity of advanced research and development available domestically. Such a technology base should consist of a broad spectrum of activities: from short-term, application-oriented development to long-range, exploratory research. Industry usually focuses on the first kind, leaving high-risk frontier research to the (often government-sponsored) programs at the universities. A dynamic, challenging scientific and technical environment, with good interactions between the parties involved, is a prerequisite for a healthy and strong technology base.

Ideally, basic knowledge and ideas should come from academia, allowing a diligent and alert industry to capitalize on them, thus ensuring further growth. There are basically three prime contributors to this technology base: universities, the federal government, and industry itself.

One might expect to find a major share of the national advanced research effort in universities. In fact, however, this source of knowledge does not seem to flow very abundantly. Inadequate financial resources at universities have caused a lag in research efforts compared to those of industry. Perceiving this situation, several universities have reacted by setting up new programs to improve their research capabilities, often in cooperation with industry.

The federal government contributes in several ways to the technology base of integrated circuits. Traditionally, the Department of Defense has been a strong sponsor of R&D, following its role as a major customer in the early years of IC's. Another source of funds for fundamental research is the National Science Foundation. However, several more indirect influences exist, such as various programs of the National Bureau of Standards.

Finally, industry itself provides most of the nation's technical know-how in integrated circuit technology. A few large industrial laboratories have long-range basic research programs, but most merchant houses have more short-term R&D activities focused on product applications.

THE ROLE OF ACADEMIC RESEARCH

A private enterprise is limited in the scope of the research projects in which it engages. In particular, its investment in research projects must have an appreciable chance to lead in a reasonable time to a useful return. A large company can afford a somewhat longer time frame than a smaller one, but in every case the scope of research is limited.

The U.S. as a nation may find it necessary to support a broader range of projects, including those of which the time frame exceeds that of industry. Such long-term projects should then be consistent with the projected needs of industry. They might be developed with government support if they are judged to be in the national interest. However, such research projects, though financed by public means, should not be directed or carried out by government. Rather, universities seem to be the proper place, assuming that mechanisms for financing and program determination can be found.

This reasoning has been articulated by industry spokesmen such as Noyce,(1) which underscores the importance with which the microelectronics industry views university research programs. Obviously, the major product of universities is well-educated students. But healthy, advanced university research is in itself considered to be of prime importance, as is also evident from the large number of interactions which exist between industry and universities.(2)

It is therefore, of interest to investigate the way U.S. universities have reacted to rapid technical change in the microelectronics field.(30) Some are concentrating substantial research programs on IC technology, often in new organizational frameworks. Initiatives of the University of California at Berkeley, the California Institute of Technology, Cornell University, the University of Minnesota, Massachusetts Institute of Technology, the University of North Carolina, and Stanford University provide good illustrations, but several others could be cited.(29)

UNIVERSITY RESEARCH PROGRAMS

University of California, Berkeley

The University of California at Berkeley has a long-standing activity in semiconductor research. Some of its achievements are:

- Computer programs for transistor design, process development, and circuit simulations.
- Novel device concepts, such as switched-capacitor filters.
- Certain new circuit designs such as analog-to-digital converters.

There are close contacts with industry, through an Industrial Liaison Program as well as through such direct contacts as grants and visiting personnel.

A unique institution at Berkeley is a centralized microelectronics technology facility that is shared among many faculty members and students from diverse disciplines. This hands-on experience is a special feature of the university's educational and research programs for graduate students.

Recently, the Governor of California proposed that the state support the Berkeley Electrical Engineering and Computer Sciences activities with a substantial amount of money.(3) The objective of this proposal is to counteract the activities of other states which are actively trying to lure high-technology companies away from their California birthplace.

A main goal of the program is to enlarge and re-equip the antiquated microelectronics technological facility with the necessary clean rooms and modern equipment for lithography, processing, testing, and characterization. This would give Berkeley the necessary facilities to carry its educational and research programs to a more sophisticated level. The intention is to pursue a policy of direct access to the facility for as many people as possible.

Funding of the subsequent research project would be done cooperatively by state and industrial sponsors. This so-called MICRO program (Microelectronics Innovation and Computer Research Operation) could substantially advance the size and level of the future microelectronics and systems activities at Berkeley.

The organizational structure is to include a board with representatives from the state, the university, and industry.

California Institute of Technology, Pasadena

At the California Institute of Technology the Computer Science Department's effort in integrated systems include the development of a structured design methodology that allows computer scientists to design integrated circuits.(4,28)

A major contribution has been the description of the designs in terms of a computer language, the Caltech Intermediate Format (CIF). This allows automatic generation of certain circuits provided the basic cell design has been previously described. Also, the total IC design can be represented in this language, providing a logical interface with the mask-making processing technologies.

Though the methodology does have a number of restrictions, it is not overconstrained. After a minimal amount of training, computer scientists have made a number of successful, rather intricate designs using the method.(5)

Caltech has organized a Silicon Structures Program in which a number of industries participate. The industries provide money and equipment and in return they are invited to send scientists to participate in the ongoing research program. This procedure is intended to serve as a major technology-transfer mechanism, provided that industries are receptive to new ideas on VLSI design.

A major educational effort has been initiated through the cooperation of Caltech with a group at the Xerox Palo Alto Research Center. A number of courses have been organized at various universities, designed for:

- Teaching the structured design methodology to computer scientists.(5)
- Making an actual design,
- Producing the actual integrated circuits,(6)
- Evaluating the performance of chips by the designers.

The designs (described in CIF) were transmitted via the ARPANET, a data network connecting many computers in the U.S., to Xerox at Palo Alto, where a computerized VLSI-implementation system was developed.(7) This system can be described roughly as follows:

- The designers can drop their designs into computer files and they can communicate with the implementation staff through an electronic-mail facility.
- A checking program ensures that the designs adhere to the CIF standards.
- An interactive layout system is used to place several designs together on one chip.
- The design data are translated onto mask tapes with prescriptions for the further fabrication process.

- Finally, documentation is prepared to be returned with the chips.

Mask making (E-beam) and chip fabrication were subcontracted. By combining various designs on one chip (multiproject chip) and various chip designs on one wafer the actual cost per design was kept low. The turnaround time for the whole operation was extremely short, about five weeks.

This successful demonstration has also led to the conclusion that the design tools and techniques can diffuse very quickly through a system like ARPANET.(8) Recently the implementation function has been taken over by the Information Sciences Institute at the University of Southern California under the sponsorship of DARPA, the Defense Advanced Research Project Agency.

Some courses using the Mead-Conway methodology are commercially available and some new ventures providing technological services along these lines have been established.

University of Minnesota, Minneapolis

The microelectronics center at the University of Minnesota originated from contacts made between university scientists and the Control Data Corporation about solid-state surface science. The university is well known in this field, and the National Science Foundation established a national center for surface-analysis techniques at the campus. It was then suggested that these novel techniques might be applicable to integrated-circuit technology.

These discussions led to a grant from Control Data Corporation, later augmented by grants from Honeywell and Sperry Univac. The purpose of the grant was to establish a basic-research program in microelectronics and information science, including surface science. The Microelectronics and Information Science Center (M.E.I.S.) was established to carry out this task.

The management structure that evolved after some time includes a management team consisting of a director with three associate directors. A group of eight representatives from sponsors and participating departments serves as an advisory board for the programs and policies.

The program is aimed at a long-range scientific effort, to be conducted over a number of years. It is intended to capitalize on the strengths present in the university. The current program focuses on four areas:

1. Materials science, where the university has a broad expertise in fields such as thin films, III-VI compounds like gallium arsenide, and material-deposition techniques.

2. Design automation, where the CALMA Division of General Electric is expected to make a substantial contribution in equipment and software.
3. Microelectronics, including the refurbishing of the present processing equipment. (There is no intention to construct a chip-manufacturing facility, since this service can be obtained from industry.)
4. Computer science, a program in which a computer-science group will establish a program in automation of software engineering in close cooperation with industry.

Programs at the various university laboratories to be funded by M.E.I.S. are expected to have an equal amount of money from federal sources available. Close cooperation with research divisions of participating industries will be maintained through personnel exchange and use of equipment.

The multiyear M.E.I.S. program will stimulate interdisciplinary work at the university. This, combined with close relationships with the participating industries, most of which are in the same geographical area, may constitute a stimulus for a new strong research activity at the University of Minnesota.

Massachusetts Institute of Technology

Recently, MIT put forward a proposal to greatly increase activities in the microelectric field. The existing Microsystems Research and Education Program covers research efforts in a wide range of projects conducted in the Artificial Intelligence Laboratory, the Center for Material Science and Engineering, the Laboratory for Computer Science, and the Research Laboratory of Electronics. The subjects include submicron structures, semiconductor materials, semiconductor processing, large-scale circuit theory, VLSI design automation, VLSI complexity theory, and IC systems architecture.

The new program would be based upon a new VLSI laboratory, featuring a computer-controlled fast-turnaround processing facility for manufacturing VLSI circuits. Additionally, the existing laboratories would be expanded to accommodate the new research programs envisaged.

The central theme of the efforts will be the integration of the various efforts needed for VLSI production. A thorough understanding of the full range of activities that constitute this process, from semiconducting materials to systems design, will be required. The program should lead to a comprehensive effort to manage the complexity of large-scale system design.

MIT proposes to finance this program by means of grants from industrial firms, which would participate in a new Microelectronics Industrial Group either as founding members or as contributors. Several exclusive benefits to the participants are proposed, including a program for visiting industrial researchers, plus workshops and seminars.

Organizationally, a Microsystems Advisory Council (composed of representatives from the sponsoring industries) will assist the Director of the Microsystems Research and Educational Program.

Microelectronics Center of North Carolina, Research Triangle Park

The officially stated goal of the Microelectronics Center of North Carolina (MCNC) is "to develop an educational and research activity in microelectronics that will establish North Carolina as a national center in this significant technology." North Carolina has been actively seeking to increase the establishment of high-technology industries in the state. This effort has already resulted in the creation of the Research Triangle Institute, a nonprofit industrial R&D organization, and the Research Triangle Park industrial area, where a number of companies have settled. In this context, an active and high-level university system is regarded as a major asset, both as a source of manpower and research and as an intellectually stimulating environment.

The MCNC will be formed by the following parties:

- Five universities: Duke University, North Carolina State University, North Carolina Agricultural and Technical State University, the University of North Carolina at Chapel Hill, and the University of North Carolina at Charlotte.
- One nonprofit institute (Research Triangle Institute).

Additionally, close cooperation will be established with the North Carolina Community College System. MCNC will have a VLSI computer-aided design facility, connected with data links to several of these institutions. It will also have a modern, fast-turnaround IC-manufacturing plant that can be used for R&D or even some pilot production.

These sophisticated facilities will be utilized by the participating institutions to improve and expand their educational and research activities in VLSI technology and systems. These activities will further be facilitated by a system of video links allowing specialized classes to be shared by the different institutions.

The main body of research will be carried out by individual faculty members and their students. However, the Center itself may also take on contracts for its own R&D activities.

It is expected that the universities will be able to educate a greater number of scientists, engineers, and technicians trained in VLSI, which may prove attractive to a broad range of high-technology companies. General Electric was the first company to announce its decision to establish a major center for VLSI in the vicinity of MCNC.

MCNC is set up as a not-for-profit corporation, separate from the normal academic administration. The Board of Directors consists of representatives from the state (one state official and three public representatives), from the universities, and the Director of MCNC, who is advised by a board of technical representatives from participating institutions.

A major contribution for getting MCNC started will come from the state; it will bear the costs for buildings and a major portion of costs for equipment and initial operations. The Center is expected to rely increasingly on contracts from government and industry in the future. An Industrial Liaison Program will be set up.

Stanford University

Stanford University's program in electrical engineering, solid-state physics, and applied physics have a high national standing. Stanford has traditionally worked closely with industry. The Industrial Affiliates Program is one example, but many other close ties exist. The presence of Stanford University has been a major factor in the development of high-technology industry in the Bay Area.

Stanford has moved toward the organization of a Center for Integrated Systems, a cooperative effort of the Department of Electrical Engineering, Computer Science, and Applied Physics to explore new possibilities generated by the VLSI technology. The stated objectives are to produce a large number of Ph.D.'s and M.S. graduates, educated in this new discipline, and to carry out fundamental research in the field of microelectronics.

A new laboratory for VLSI technology will be set up. It will be a highly computer-controlled, fast-turnaround IC-processing facility which will provide faculty and students with IC's according to their design. Close cooperation between IC designers and technologists will be established.

Funds for the plan will be shared by the university, government, and industry. A novel scheme for industrial sponsors has been set up. They will make a three-year commitment for $250,000 annually, money that will be used chiefly for a new building to house the facilities. The operating expenses are expected to be shared approximately equally by government and industrial research contracts. In the government sector, the National Institutes of Health are

TECHNOLOGY BASE   137

expected to be a continuing supporter of Stanford's microelectronics program.

Organizationally, an Executive Committee will set the major policies for the Center. The policies will be implemented by two directors, responsible respectively for relations with industry, including education, and for relations with government and the management of the fast-turnaround manufacturing facility.

The actual research will be done by individual faculty members and their students in various participating departments, making use of the new facility. Some of the projects will be carried out in the framework of larger contracts with government or industry and will then be coordinated by the directors. Other projects will be handled individually by faculty members.

Policies for the relations with the sponsors have to be determined. The academic nature of the research precludes any proprietary research being conducted. However, sponsors will get some preferential treatment. They will be represented in an Advisory Committee and their researchers may be invited to participate in some of the Center's activities. An ongoing educational program (e.g., via videotapes) for the participating industries will be among the services provided.

Motivations

In reviewing activities at universities striving to improve their strength in integrated circuits research, several different motivations are distinguishable in addition to a common concern regarding education in this discipline.

- MIT and Stanford intend to secure a leading position in "VLSI science" by addressing major current problems in the field. Both envisage special organizations to obtain that goal. The MIT program will probably have a somewhat broader scope than the one at Stanford, which in practice, is closely tied to its NIH contracts. Both universities intend to set up a fast-turnaround manufacturing facility. This idea is sometimes disputed: the high costs for capital investment, depreciation, and operation will be a heavy burden on budgets, whereas much of the work could be subcontracted with industrial facilities. On the other hand, device and systems work will probably be better served with a nearby facility fully committed to research goals.
- Berkeley's goal is to be a center of excellence in VLSI techniques. As a state university it has a general task of supporting the California industry. Its program

seems to be more student-oriented than those of MIT and Stanford. Typically, its new technological facility will be highly accessible to staff and students, rather than being a computerized, automated arrangement. Nor is a special organizational setup considered: the current cooperation between the Electrical Engineering and the Computer Sciences groups is seen as adequate.

- More regional models are provided by the University of Minnesota and the North Carolina schools. The former, with its M.E.I.S. organization, is an example of regional cooperation between university and industry, which may be expanded on a national level. Local industry can provide the processing facilities; the organizational arrangement is thus rather simple. In North Carolina, MCNC is primarily meant to be a service center for the universities. This would boost academic educational and research programs which would in turn lead to an improved technological infrastructure for the state. Seed money, donated by the state, will be used to set up a design and manufacturing capability since industrial capabilities are not readily available.

- Activities of the California Institute of Technology are rather restricted, focused on the important problem of VLSI design methodology. The industrial contacts serve also as a technology-transfer mechanism. The absence of an equally strong IC production activity may be one reason that some in the industry view this work with some suspicion. The more recent educational efforts, including fast realization, undertaken in cooperation with Xerox Laboratories, have been a very important step to the broader diffusion of the basic Caltech ideas.

INDUSTRY-UNIVERSITY RELATIONS

The academic world is responding to the challenge of microelectronics, but this reaction seems rather slow: only recently have some major projects been initiated, whereas the specific needs of future microelectronics could have been determined many years ago.

One should remember, however, that microelectronics is not a defined academic discipline. It can take the form of VLSI-systems design, focus more on devices, or lean more towards computer science. In any case, it requires the cooperation of several departments, such as electrical engineering, computer sciences, and applied physics and chemistry. Establishing such cooperation and obtaining funds for purchasing and running the necessary expensive equipment is a major undertaking. However, the results can be rewarding. The

NSF funding of Cornell as a National Center is a good example of the possibilities offered by a combined effort.

Is the current effort at the universities enough? It is difficult to evaluate, but the impression in industry circles is definitely negative. Though some major universities are building up their capabilities, the situation at the second-tier schools, especially their engineering departments, is unsatisfactory in terms of faculty and research equipment. A critical mass of activities and a certain level of sophistication are essential both for education and for research on real problems which are of interest to industry.

Raising the necessary funds is a major problems for the schools. For the time being, industry contributes a rather small fraction to university research budgets, probably less than 5 percent.(2) It has been argued that in the current economic climate and tax environment it is not realistic to expect a change in this situation.(9) However, as the federal government cuts back its support of academic research, expectations are aroused for business to assume some portion of sponsorship. This situation has raised the issue of whether new tax laws giving companies tax credits for corporate contributions to a university would be desirable.

In those cases in which a major contribution is obtained from private enterprise, as with the Stanford Center for Integrated Structures, the question arises about which benefits the sponsor is entitled to. It seems reasonable to expect some, since otherwise the funds could have been applied to pulling together a small in-house research team to do the work.

Universities, on the other hand, stress freedom of research and publication. Research that is carried out will not be of a proprietary nature; any ensuing patents will be held by the university or its faculty. However, universities recognize the need to devise some measures to reward the sponsor's unique position. For instance, Gray(10) notes that MIT prefers to take its own patents, which will be licensed to the sponsor on a royalty-free nonexclusive basis. However, MIT accepts limited-time exclusive patent agreements when this arrangement is necessary for the sponsor. Similarly, though MIT stresses that its research results are in the public domain, it is willing to delay publication for a reasonable period to safeguard a sponsor's interests. Also at Stanford's Center for Integrated Circuits, ways are considered to give sponsors a favored position. For example, in-house training programs will be established, with consideration of the possibility of choosing visiting industrial researchers preferentially from laboratories of a sponsor.

Of course a certain unrest vis-à-vis the traditional openness of the university research begins to develop. (See, for instance, Gray(10) and Greenberg(11) for conflicting views, and Peters(2) for a more general perspective.) Examples of agreements discussed above concern very prestigious universities which have a strong bargaining position. Lesser-known institutions may be obliged to accept much less favorable agreements. Nevertheless, it will be interesting to see which solutions prove satisfactory to the parties involved, including outside interests such as nonparticipating industries.

In the past few years suggestions have been made to draft legislation that would permit companies to sponsor university research while obtaining tax credits. It is argued that this could then be a major new source of funding. Indeed, if the tax credit amounted to a maximum of 10 percent of industrial R&D expenditures, it would total roughly $60 million for the merchant and captive semiconductor industry together. This money would then probably be used mainly to sponsor applied sciences and engineering, which many in the industry feel have been short-changed in recent years.

What would be the effect of such a new source of funding, distributed by individual merchant houses and captive producers? One can speculate that the top universities would be able to carry out still more ambitious work than today. But since they are consciously limiting their size to maintain quality, probably the second-tier universities would be benefiting the most. This effect might be strengthened by the fact that many industries would exploit contacts with nearby universities.

Even so, a certain danger may exist for an overlap of activities, since several industries would establish similar research programs at different universities. Some coordinating mechanism, either between industries or between the universities, might help to prevent this. For instance, an organization such as the Semiconductor Industry Association could try to obtain a coordination of industrial sponsors. An effort to establish a coherent research program for all industry, however, might lead to a concentration of grants in a few centers of excellence and a continued underfunding, certainly in terms of importance of research projects, of the second-tier universities.

However, several representatives from the IC industry would probably not be in favor of a National Science Foundation system of program determination by means of proposals and peer reviews. This process is regarded as a self-perpetuating mode of operation which would not keep pace with the quickly changing technology needed by industry. Industry would require some mechanism to ensure influence in program determination and to present necessary feedback after completion of the research. Nevertheless, it seems reasonable that academic research would be relatively free from the fluctuations in budgets, goals, and priorities that often occur in an industrial environment.

The issue of proprietary rights would probably not be a major problem in an extended cooperation between IC industry and universities. Usually a rather clear line may be drawn between the public and the private domain. Industry should acknowledge that those research projects which are essential for their product lines should be carried out in-house: they should not regard a university as a source of cheap product development. On the other hand, a university should, for example, refrain from the fabrication of commercial circuits. Regarding patents, experience shows that usually a satisfactory solution can be found.(2) Also, some delay in submission for publication is often accepted by the university.

In any case, a number of new and intricate questions arise if the IC industry becomes a major funding source of academic research on semiconductors. It seems clear that industry should act carefully in exercising influence which may gradually change the focus of academic research while not altering its traditional purview. To this end a common approach by industry would be in order to ensure a satisfactory combination of a coherent research program for all and for more specific programs for particular industries. The recent establishment of the Semiconductor Research Cooperative by the Semiconductor Industry Association is a first step in this direction.

THE FEDERAL GOVERNMENT

The contributions of the federal government to the national technology base occurs primarily through research contracts to universities; the National Science Foundation, the National Institutes of Health, and the Department of Defense are prime examples. A more direct contribution is made via the work at the National Bureau of Standards, which is an agency of the Department of Commerce.

The National Science Foundation; the Cornell Facility

The National Science Foundation is an independent agency. Its director reports directly to the President. With a current budget amounting to just under $1 billion, it finances university programs in basic science and engineering. Additionally, an appreciable amount of money is spent in science education. Scientists submit unsolicited proposals, which are subjected to peer review. Annually, 26,000 proposals are received of which 12,500 are awarded to 22,000 performers. Some 50,000 persons act as reviewers. The twenty-five advisory committees have a major influence on the priorities within the total budget, although NSF as such does not have a "national strategy" per se.

Numerous research grants are related to the microelectronics field. They include such areas as solid-state physics, materials science, and device physics. In addition to this broad spectrum of research carried out at many universities, a concentrated program on very small-scale technologies has been initiated. Such a program is aimed at closing the gap that has been widening between the facilities in industrial research institutes and those at the universities. This program led to the establishment of the National Research and Resource Facility for Submicron Structures at Cornell University in 1977.(12)

The form of a national facility has been chosen because of the high cost of doing first-class research in submicron structures. The necessary equipment is expensive and entails substantial operating costs. Rather than scattering the research through small grants over a number of universities, the NSF has chosen to make a long-term commitment of nearly $7 million over five years at one place. Cornell was selected as the location for the facility among seventeen proposals. In addition, the university has provided funds for a new laboratory with the necessary provisions for such problems as vibration and electromagnetic disturbances.

The facility is headed by a director, assisted by two associate directors. The director is advised by:

- A Policy Board, consisting of members from academia and industry.
- A Program Committee, in which NSF, industry, and universities are represented.
- A User Group, representing the various users of the facility.

The Cornell facility has two tasks: it operates as a well-equipped laboratory for visiting university researchers, including several Cornell faculty, and it conducts a research program in microscience on its own.

The program focuses on experimental techniques to produce and investigate structures with a size of 1 micron and below. Currently, much of the work is below 0.1 micron, with the smallest structures of 15 Ångstrom. The available resources include equipment for the generation and replication of patterns, the preparation of materials and thin layers, semiconductor-type process technology, and analysis of devices and structures. Additionally, computer facilities for design purposes are available.

The facility has several continuing projects aimed at developing experimental techniques and enhancing present capabilities. The research programs by the Cornell faculty encompass eight different departments. Currently, thirteen other universities use the facility for twenty programs. The main body of research is in areas such as electronic devices and integrated optics ("nano electronics"); some molecular research is also proceeding.

An industrial-affiliates program has been established with a membership of twenty-five companies. They are kept informed of the ongoing research and are invited for an annual meeting. Though several industries have given research grants to the projects, industrial researchers have made little use of the facility.

Other funding is obtained from governmental agencies other than NSF, notably the Department of Defense: several projects for the VHSIC program are carried out at the Cornell facility.

National Institutes of Health

The National Institutes of Health award research grants for basic work in many areas of the medical field. Usually research proposals are initiated by scientific investigators. NIH submit proposals to a very thorough peer-review procedure: at special meetings the reviewers discuss the proposals in depth. If a favorable conclusion is reached, the Institutes determine the relevance of the proposal to their mission and assign it to one of the ongoing programs. In a few cases, NIH take the initiative by requesting applications.

The vast majority of NIH contracts go to universities or to institutions that are affiliated with universities. Industrial-research laboratories play only a small role, though sometimes they can bring valuable technological know-how to the projects.

A couple of grants are awarded that are relevant to the technology base in microelectronics. These grants total less than $5 million. A major beneficiary is Stanford University, although a program has been initiated at Case Western Reserve University.

At Stanford, implantable sensors(13) are being developed that can measure a variety of parameters in the human body through an extended period. Suitable telemetry devices form a part of this development. Major problems are caused by the hostile environment, which puts stringent requirements on packaging techniques. Leakage and zero drifts are common problems that have to be solved, for example by the use of suitable digital-circuit techniques.

Case Western's special knowledge is in the use of field-effect transistors as sensors in a fluid environment. An application of a FET device for accurately sensing the pH has been industrialized. Similar techniques are used in the construction of a potassium sensor. Future applications may include the detection of enzymes over long periods in the human body.

Some smaller programs are also funded, for example at the Universities of Utah, Michigan, and Arizona. Some work on VLSI-design techniques is sponsored at the University of Washington and at Washington University, St. Louis.

Department of Defense

The Department of Defense is a major sponsor for R&D in electronics and computer science.(14) The research part of its Research, Development, Test and Engineering program contains nearly $50 million for electronics and $35 million for mathematics and computer sciences. This money is allocated roughly equally among DoD labs, industry, and universities. It covers many subjects, for instance gallium-arsenide integrated circuits, acousto-optics, submicron structures, and high-frequency transistors.

In 1979 the six-year very high-speed integrated circuits program (VHSIC) (15,16,30) was funded with over $200 million. Coordinated by the DoD, specific contracts are awarded by the three armed services. The aim of the program is to provide by the mid-1980s a number of military systems in which high-speed, real-time data processing is essential. There will be a large range of applications, including radar, sonar, missile control, and electronic intelligence.(27)

Together with the design of such systems, suitable integrated circuits have to be provided. They will have a greater complexity and speed than the IC's available today. In addition, a low power consumption is desirable and some specific military requirements, like radiation resistance, must be met. These general signal-processing chips will be programmable for the specific applications desired.

After an initial phase 0, contracts have been awarded for phase 1 to a number of systems companies and IC manufacturers.(25) This entails the development of IC's with 1.25-micron line width and the technology for 0.5-0.75 micron-line widths. In phase 2, phase 1's 1.25-micron circuits will be applied in demonstration systems and new circuits will be developed in the 0.5-0.75 micron technology.

Concurrently with phases 1 and 2, a phase 3(17) will fund a number of more specific subjects which support the main line. Examples are high-resolution lithographic equipment, IC packaging, fault-tolerant design concepts, and test methods.

The VHSIC program has won mixed reviews from the industry. Some companies criticized Defense for draining professional talent from the development of commercial IC's, thus weakening the industry's international competitive position. The potential commercial spin-off of the program, primarily in the high-resolution technology, is not thought to be worth the effort. Such technologies are being developed by industry anyway, and a better way of spending money would be in research on VLSI systems design.

Another criticism has been the absence of very specific circuit goals, to be realized in a competitive environment. Under present agreements the participating industries would not be compelled to spend their best resources on the program.

The security regulations imposed on the VHSIC project have begun to arouse concern at participating universities and industries. For example, they are expected to exclude foreigners from working on the project and to restrict the exchange of scientific information. This restriction poses considerable practical problems at many graduate schools where about half the students are foreign.(18) The principal problem lies in the restriction of the information flow, which is in contrast to the normal academic practice.(19) In February 1981, the presidents of six major universities highlighted this issue by expressing their concern about the current interpretation of the International Traffic in Arms Regulation (ITAR) applied to the VHSIC program, and asked the secretaries of the departments involved to clarify the regulation in a manner compatible with traditional academic freedom.(20)

National Bureau of Standards (NBS)

The National Bureau of Standards is an agency of the U.S. Department of Commerce. In its National Engineering Laboratory, the Center for Electronics and Electrical Engineering is responsible for research related to integrated circuits. The Center has organized a Semiconductor Technology Program which covers a rather broad range of activities:(21)

- Materials characterization.
- Device characterization and testing.
- Process control.
- Test structure and methods.
- Standards, consultation, and dissemination.

Generally speaking, NBS is mainly concerned with developing appropriate measurement technologies. Some examples of the problems that have been tackled by NBS are:(22,23)

- The measurement of line widths on a microcircuit with simple optical methods. A standard pattern has been developed that can be reproduced on a chip. In conjunction, microscopic techniques were adapted which allow the measurement of 0.5-micron lines with accuracy within 10 percent.
- Test structures have been designed which can be produced on a slice together with the IC's. Such test structures allow the measurement of specific parameters after processing has been completed.
- A program is underway to discover causation of the various defects in the crystalline silicon and to develop techniques to measure these defects.
- Computer models of processes and devices are under study. These will allow the investigation of the effects which occur when the device dimensions are reduced further.
- The properties of hermetically sealed encapsulations have been investigated.

The NBS work is usually initiated after informal contacts with industry, professional societies, trade organizations, and academia have identified existing problems. Such problems often occur between companies, for instance on whether a silicon wafer conforms to a specified surface resistance. Also, in the manufacturing process standardized measurements are useful, for example in determining physical properties with standard test structures. The results of NBS investigations are published(23) and communicated to the industry. A special quarterly publication(24) is issued, and conferences and workshops are organized.

These activities often lead to measurement methods which are widely used. They may appreciably reduce disputes about specifications, aid process control, eliminate sources of faults in production, or help in understanding underlying physical phenomena. Some of the methods are introduced later in formal standards.

Thus the NBS work may result in significant economic returns to the industry at large. A study of cost-benefit effects has been carried out by the Charles River Associates.(21) This report studies, among other approaches, the cases of three semiconductor-research projects and their impact on the industry, qualitatively as well as quantitatively. It shows that a high return on investment has resulted from those NBS programs.

The current annual expenditures of the Semiconductor Technology Program amount to about $4 million. The major part is funded by the Department of Commerce. About one-fifth of the program consists of contract work for other agencies, mostly the Department of Defense.

The program is overseen by an Evaluation Panel consisting of experts from industry and universities. Once a year this panel meets to discuss the current and planned activities in detail.

The Semiconductor Technology Program of the NBS is probably unique in the world, in the sense that a government agency provides the IC industry with essential measurement methods and tools. It will be continued with an emphasis on the specific requirements for VLSI, including principles of high-speed digital testing methods and measurement of power dissipation in VLSI chips.

INDUSTRIAL R&D

The major component of the U.S. technology base in integrated circuits is composed of private-sector R&D. The merchant industry spends annually on the order of $400 million (as of 1980) on industrial R&D. Figures for the captive industry are unknown, since their IC research as a part of a larger R&D activity is not reported separately.

Several captive producers have industrial-research laboratories which are among the most important contributors in the field. The strongest position is held by IBM Laboratories and Bell Laboratories, but other companies such as General Electric, Hewlett-Packard, and Honeywell have considerable activities in industrial research on integrated circuits.

One indication of the strength of industrial R&D is the number of patents on IC's obtained in the ten-year period 1970-79. This amounts to 1,500 for five representative merchant houses and 2,500 for IBM and Bell Labs together. For comparison, five representative Japanese manufacturers obtained 1,200 patents.

In the U.S., much company-specific know-how spreads rather quickly, because of several phenomena:

- The transfer of people between companies, which is spurred by the intense competitive climate of the industry.
- The establishment of joint ventures, often for a limited period, by several firms.
- The practice of second-sourcing circuits, which entails sharing design knowledge.
- The existence of various cross-licensing agreements relating to processes or designs.

As a consequence, lead times regarding new developments are relatively short. In the IC industry lead times in excess of more than one year are usually considered unrealistic: it is expected that competition will quickly catch up through these mechanisms. Compare also Sze's observation(26) that on average a two-year delay exists between publication of R&D results and the business activity based on those results.

However, this type of industrial research is highly fragmented over many companies. This does not hold for IBM and Bell Laboratories, but in turn their work does not lead directly to products on the merchant IC market. The U.S. merchant houses are therefore not in a particularly strong position to counter the more concentrated R&D efforts of the major competitor, Japan. In that country industrial R&D expenditures by the six major companies, augmented with some government subsidies, are roughly equal to those of the American merchant sector.

Such considerations have led to industry proposals to join R&D forces. The Semiconductor Industry Association has established the Semiconductor Research Cooperative. A group of computer and IC manufacturers are setting up a common company, tentatively called Microelectronics & Computer Technology Corporation, to carry out advanced VLSI and computer research.

REFERENCES

1. Noyce, R.N., Private communication, 1981. See also his discussion on "Capture Ratios," in Davis (Chapter 2, Ref. 5); and private communication.

2. Peters, L.S., "Current U.S. University Industry Research Connections," Center for Science and Technology Policy, New York University, publication in preparation.

3. "Governor Brown Boosts Microelectronics," Science 211 (Feb. 13, 1981), 688.

4. Mead, C., and Conway, L., Introduction to VLSI Systems, (Reading, MA: Addison Wesley, 1980).

5. Clark, J., "A VLSI Geometry Processor for Graphics," Computer, July 1980, p. 59.

6. Hon, R.W., and Sequin, C.H., A Guide to LSI Implementation, Second Edition, Xerox Palo Alto Research Center, SSL-79-7 (1980).

7. Conway, L., Bell, A.; and Newell, M.E., MPC '79, Lamda, Second Quarter 1980, p. 10.

8. Conway, L., Lamda, Fourth Quarter 1980, p. 65.

9. Pake, G., "Industry-University Interactions," Physics Today, January 1981, p. 44.

10. Gray, P.E., "MIT Wants Closer Ties with Business," New York Times, Sept. 27, 1981.

11. Greenberg, D., "Academic Science for Sale," New Scientist, July 16, 1981, p. 174.

12. Wolf, E.D., "The National Submicron Facility," Physics Today, November 1979, p. 34.

13. Barth, P.W., "Silicon Sensors Meet Integrated Circuits," IEEE Spectrum, September, 1981, p. 33. See also Middelhoek, S., et al., "Microprocessors Get Integrated Sensors," IEEE Spectrum, February 1980, p. 42.

14. DoD Basic Research, DoD publication, August 1980.

15. Sumney, L.W., "The United States Department of Defense Program on Very High Speed Integrated Circuits (VHSIC)." Thirteenth Asilomar Conference on Circuits, Systems and Computers, Pacific Grove, CA, November 5-7, 1980.

16. Sumney, L.W., "VLSI with a Vengeance," IEEE Spectrum, April 1980, p. 24.

17. Sumney, L.W., "Pipelining Innovation into VHSIC: The Phase-3 Concept," Military Electronics/Countermeasures, May 1980, p. 50.

18. Chesser, T.V., "Foreigners Snap Up the High Tech Jobs," New York Times, July 5, 1981, p. F13.

19. Dickson, D., "Academe Ponders Defense Curbs on Research," Science and Government Report, March 1981, p. 5.

20. Washington Post, April 9, 1981.

21. Charles River Associates, Inc., Productivity Impact of R&D Laboratories: The National Bureau of Standards' Semiconductor Technology Program, Boston, 1981.

22. Bullis, W.M., Advancement of Reliability, Processing and Automation for Integrated Circuits With the National Bureau of Standards. Final Report, March, 1981. NBSIR 81-2224.

23. Semiconductor Measurement Technology. NBS List of Publications 72, March, 1981.

24. Bullis, W.M., (ed.), Semiconductor Technology Program Progress Brief. NBSIR publication.

25. Posa, J.G., "VHSIC Competitors Reveal Phase-1 Design," International Electronics, Sept. 22, 1981, p. 89.

26. Sze, S.M., "Semiconductor Device Development in the 1970's and 1980's - A Perspective," Proceedings IEEE 69 (1981), 1121.

27. "VHSIC Brassboards Hint at Potential," Defense Electronics, September 1981, p. 35.

28. Marshall, M., et al., "The 1981 Achievement Award," Electronics, Oct. 20, 1981, p. 103.

29. Main, J., "Why Engineering Deans Worry a Lot," Fortune, Jan. 14, 1982, p. 84.

30. University/Industry/Government Microelectronics. Symposium, Starkville, Miss. IEEE, 1981.

9
The Manpower Problem

The entire electronics industry is seriously concerned about the availability of well-trained people, primarily those with backgrounds in engineering and computer science. This concern is shared by many in academic circles who have witnessed recent enrollment declines in selected areas. For example, many feel that engineering education has long been neglected and that facilities at engineering schools are inadequate.(1,2)

The American Electronics Association (AEA) created a "Blue Ribbon Committee" on engineering education consisting of representatives from industry and academia.(3) The function of this group was to determine the nature of the manpower shortage and to recommend a plan of action. The committee chairman has stated that the shortages of engineers "pose a serious threat to national security . . . to our economy . . . and to the continued vitality of electronics industries, where the lack of electronic and computer science engineers may be the single most important factor limiting factor growth."(3)

Let us briefly examine the existing data on the expected future supply and demand of engineers and discuss some of the commonly observed deficiencies in the educational system.

DATA

A limited amount of data is available on the need for technical people in the industry. In 1980 the AEA conducted a manpower survey on needs for the entire U.S. electronics industry. An estimated 39 percent of the industry replied, measured according to sales volume. The survey participants were asked to project the number of jobs in their companies for 1981, 1983, and 1985 in a variety of technical areas. The overall results indicate(3) that over a five-year period the following compounded growth rates are expected:

Electronics-Electrical Engineer	12%
Electronic Engineer/Technologist	16%
Software Engineer	16%
Analyst Programmer	18%
Other Computer Professionals	18%

Data such as these should be regarded with some reservation. It has often been difficult for the electronics industry to practice proper manpower planning, in view of the fast-changing technologies and markets. Other factors must be taken into consideration, such as the general economic situation, a possible increase in the productivity of professionals in this field, and the ability of people to switch to other types of jobs.

However, the actual number of individuals needed will exceed the percentages mentioned. Many presently employed professionals will leave the industry during this period, either by retirement or by transferring to jobs outside the industry. Additionally, a number of senior professionals will assume managerial positions.

Another consequence of the growth of the electronics industry is the expansion of electronics into fields that have been hitherto nonelectronic in nature or even nonexistent. Such industry segments have not been included in the AEA survey. Nevertheless the electronics content in the job requirement for someone like a garage mechanic will increase considerably. Also, the introduction of electronics, for instance in a small-business environment, will require many new professional activities such as the provision of application software. These are tasks which are quite often suitable for small companies.

In any case, one may expect that more extensive and detailed studies would reveal a considerable increase in demand for professionals in electronics and computer science in the coming decade.

In terms of the more narrowly defined semiconductor industry, some estimates are available from the Semiconductor Industry Association.(4) In 1979 the U.S. merchant semiconductor firms employed approximately 230,000 people throughout the world; half of them within the United States. Employment in these merchant houses has increased at an annual rate of 10 percent. Future extrapolation of this growth rate results in a 1985 estimate of worldwide employment projections in the U.S. merchant semiconductor industry of 390,000 people, and estimates for 1990 are 620,000. Again, several factors may disturb this picture. In particular, the early-1980s recession in IC trade will have a profound effect. Furthermore, the need for production personnel will grow less strongly because of increased mechanization. However, SIA expects in particular a substantial increase in demand for engineers and other technical personnel.

The Association estimates that, on average, to support each additional $1 million of sales, one additional electrical engineer is required. When we consider the rather phenomenal increase in sales revenues by many of the firms in recent years (typically 20 to 30 percent annually), the tremendous increase in demand for electronics professionals is obvious.

The SIA is worried that the industry's competitive ability will be seriously impaired if the tremendous projected growth in personnel cannot be realized. In addition, the problem may become more acute as devices become more complex: in particular, designing increasingly complex VLSI devices will require much more labor input.(4)

Nevertheless, estimates such as those made by SIA are only concerned with very broad categories. A much more detailed investigation of present and future employment, in terms of various types of scientists, engineers, and technical personnel, might reveal a much better insight into problem areas that have to be anticipated. Personnel projections always imply certain assumptions about the economic development in the industry. In addition to this uncertain factor, reasonable and different assumptions must be made regarding the effects of increasing productivity(5) for the various personnel sectors.

We should note also that current labor statistics do not take into account the changing character of jobs in the electronics industry.(12) For instance, designing electronic equipment increasingly requires a broad knowledge of hardware and software techniques. Systems architecture, computer-aided design, simulation, testing, chip design, and process technology should all be among the designer's professional skills. Comparatively few people are available nowadays who can muster this unusual combination, and few schools can provide the necessary training. Careful consideration of possible interdisciplinary approaches might be justified in both academia and industry.

Data are also scarce on the actual number of engineers and computer scientists whom the universities and technical colleges may graduate in the future. The number of electrical engineering graduates, which reached a minimum in the mid-1970's, is gradually increasing (see table 9.1). The same holds for computer engineers. Apparently the interest in an engineering education among students had waned during the earlier part of the decade. This interest is now growing again in view of attractive salaries and opportunities in the

Table 9.1. Annual Electrical Engineering Graduates
(Total)
U.S. and Japan

YEAR	U.S.	JAPAN
1969	16,282	11,848
1970	16,844	13,889
1971	17,403	15,165
1972	17,632	16,052
1973	16,815	17,345
1974	15,749	17,419
1975	14,537	18,040
1976	14,380	18,258
1977	14,085	19,257
1978	14,701	20,126
1979	16,093	21,435

Source: Semiconductor Industry Association, "The International Microelectronic Challenge," May 1981.

job market. Several universities and technical colleges are already reporting capacity problems.

CONSEQUENCES FOR EDUCATION

It is the consensus that the need for electrical engineers and computer scientists will increase strongly in the U.S. but that the supply of young graduates will in all likelihood be insufficient.(6) This condition represents a serious threat to the future competitive position of the U.S. industry. In contrast, for example, in Japan the number of engineering graduates shows a healthy increase. The number of electrical engineers graduated there is even larger than in the U.S. (see table 9.1), whereas Japan's population is only half that of the U.S.

Major problems of capacity are evident at universities(6,10) which are not capable of enrolling the desired number of qualified applicants:

- Most of the prestigious universities limit the number of students to reflect their standards of excellence,
- The "second tier" schools are often inadequately staffed, especially regarding their engineering faculty. A factor

here is the great difference between salaries in industry and those in academia.(9)
• Equipment at universities is often antiquated. Some top universities are moving to attract funds which will dramatically improve their equipment, but many others lack the necessary resources.

In addition to this less than hopeful situation, American students are often reluctant to engage in graduate and postgraduate education. Highly competitive salaries offered by industry,(8) the substantial financial sacrifices for pursuing graduate work, and diminished prospects for financial support from fellowships resulting from government cutbacks combine to reinforce the disincentive for advanced graduate study.

A remarkable occurrence has taken place in recent times in that foreign students now compose a large percentage of the student body at graduate schools. For example, of all 1980 engineering Ph.D. degrees conferred, more than 46 percent were granted to foreign students.(7) In fact, many remain in the U.S. after graduating and compete for top technical jobs. Though this influx certainly eases industry's problems, it results in a certain dependence on the competitive attraction of the labor market abroad. Moreover, it puts a certain stress on relations with many developing countries from which many of these students have come but to which they do not return to participate in the development of their national economies.

Another problem lies in the faculty staffing of universities in the future. In the past, faculty were recruited from Ph.D.'s who had a great interest in basic science. Current developments leave few Ph.D.'s available for university positions, leading to an increased shortage of teaching and research staff.

The question may be posed whether industry itself pays sufficient attention to the technical training and retraining of existing staff. Though this seems an obvious and necessary action in a rapidly changing technical environment, many problems are known to exist in relation to in-house training. Several companies such as Bell Laboratories and Hewlett-Packard have excellent training programs, but many others are less active. The possibility of better training to achieve improved utilization of the existing workforce is a consideration which deserves greater attention.

At the federal level, there seems to be relatively little interest in the nationwide manpower problems. Unlike in most European countries, the U.S. federal government is not constitutionally responsible for the nation's higher educational system. Thus, concerns such as adapting the educational facilities to the nation's needs receive little federal attention. It is interesting to note that the AEA Blue Ribbon Committee does not include a current representative from government. One might ask whether there are appropriate mechanisms for a more active federal policy in an internationally competitive environment where a high-level technical competence may be essential for the nation's economic and strategic position in the future.

Also industry could, in a variety of ways, help universities in carrying out their task. Donations of equipment or funds and making industry researchers available as visiting professors are among the possibilities.(10,11)

More detailed studies could, for example, examine features of the European and Japanese educational planning systems that might be relevant models. These might suggest ways of adapting the current practice in the U.S. while keeping intact the virtues of the present decentralized educational system.

REFERENCES

1. Stockton, W., "The Technology Race, America's Struggle to Stay Ahead," New York Times Magazine, June 26, 1981.

2. Pettit, J.M., "Engineering Supply and Demand in the U.S. - A Current Crisis," paper presented at the VII Congreso de la Academia Nacional de Ingenieria, Oaxaca, Mexico, 1981. Provided by the National Academy of Engineering.

3. American Electronics Association, "Technical Employment Projections 1981-83-85," May 1981.

4. SIA, The International Microelectronic Challenge. The American Response by the Industry, the Universities and the Government. (Cupertino, CA: Semiconductor Industry Association, 1981).

5. Finan, W., "The Semiconductor Industry's Record on Productivity," Technecon Analytic Research.

6. Khane, S., "A Crisis in Electrical Engineering Manpower," IEEE Spectrum, June 1981, p. 10.

7. Chesser, T.V., "Foreigners Snap Up the High Tech Jobs," New York Times, July 5, 1981, p. F13.

8. Abelson, P.H., "Industrial Recruiting On Campus," Science 213 (1981), p. 1455.

9. Main, J., "Why Engineering Deans Worry a Lot," Fortune, Jan. 11, 1982, p. 84.

10. Khane, S., "Cracks in the Ivory Tower," IEEE Spectrum, March 1982, p. 69.

11. "Hi-tech Companies Help Breed Engineers," Business Week, March 1, 1982, p. 25.

12. University/Industry/Government Microelectronics. Symposium, Starkville, Miss. IEEE, 1981.

10
Microelectronics and Employment

In the preceding chapters microelectronics has been depicted as a technology-driven phenomenon. It is quickly penetrating society because it offers inexpensive entertainment, cheap computing power, or economically justifiable means of increasing productivity - just to name a few attractive features.

Many questions of a nontechnical nature arise in the course of this process of introducing a new technology. For instance, is there a copyright for software programs? What safeguards protect private data in computer data banks? Should transborder data traffic be regulated? Such questions are being answered differently in different countries. The extent of governmental influence depends strongly on the beliefs that a nation has regarding the proper role of its public power in its particular society. Consequently, major differences in legislation and public policy exist among the leading industrialized countries.

In the broad spectrum of social issues accompanying microelectronics, let us focus on employment effects. The question whether unemployment will be caused by a strong increase in productivity resulting from the application of microelectronic technology is increasingly debated. The discussion is particularly intense in Western Europe, whereas in the U.S. and Japan there is much less concern. In the U.S. a healthy general development of the economy is seen as a stimulus for private industry to create new jobs. The assumption is that the flexibility of the private sector would more or less automatically lead to a reasonable level of employment. In Japan a large degree of consensus exists that high-level industrialization and automation is the only way to survive as a modern welfare state. This, and the traditional full-employment policy of major companies in Japan, makes the unemployment question a non-issue there. Recently, however, a growing resistance to the introduction of robots in the automotive industry(20) has been reported within the Japanese labor unions, see also government statistics reported by the Ministry of Labor.(21)

A SURVEY OF THE LITERATURE

A review of European studies reveals that the effects of a new technology on employment cannot easily be quantified. A study of some case histories by Rothwell and Zegveld(1) shows that many other factors often obfuscate the influence of the introduction of a new technology. There are several factors which have strongly influenced employment situations but which bear little relation to technological progress:

- In the industrialized countries long-term trends in different economic sectors are evident. For instance, the agricultural labor force has gradually diminished in recent decades. This loss has been compensated for by a growth in industry (until recently when the "jobless growth" phenomenon became apparent) and in the service sector (including the government).
- International competition of low-wage countries has strongly increased in several industrial sectors. It has forced firms to put strong emphasis on reducing production costs and initiating manufacturing operations in low-wage countries. The textile industry is a good example.
- International demand patterns have sometimes changed drastically, leaving a worldwide industrial overcapacity as in the shipbuilding industry.
- International trade factors such as tariff barriers, dumping practices, or export subsidies can strongly influence the performance of an industrial sector.

Consequently, any prospective quantitative study of the influence of microelectronics on employment should be viewed against a highly complex and rapidly changing social and economic background. In spite of the many methodological difficulties, various quantitative studies have been conducted for several economic sectors. A number of such studies have been collected and compared by the West German Ministry for Science and Technology, BMFT.(2) Generally, such studies include a scenario illustrating how the new technology is introduced over the course of time. This requires assumptions relating to the resulting productivity, competition, and expansion of the market. Also, estimates must be made for the capital needed to introduce the technology and for the availability of such funds. With this scenario, the expected effects on employment can be categorized and quantitatively estimated.

Such procedures seem feasible within one sector. However, the aggregation of many sectors into a single nationwide picture is as yet impossible. This is certainly true of an all-pervasive technology such as microelectronics. Too many unknown and dynamic interactions are present, specifically regarding the creation of new industrial activities and services surrounding a changing economic sector.

The BMFT analysis shows that the various methods employed lead to strongly differing results. Perhaps the only common conclusion is that many jobs will be affected in some way: a number of jobs will become obsolete, new ones will be created, and the content of many will change. The service sector seems to be particularly affected since numerous tasks here can be automated. This holds special significance, since historically many new jobs have been created in this sector, compensating in particular for job losses in the agricultural sector.

Other studies of the potential employment consequences have been undertaken by European governments, by trade unions, and by universities. One early and highly regarded study was conducted for the French government by Nora and Minc.(3) The authors outline the forthcoming merger of computer technology and telecommunications technology, for which they coined the word "telematique," - telematics - which will have the effect of accelerating the development toward an "information society."(4) They envisage that this new technology will be so widely used by the state, business, and services that it will grow into a commodity similar to electricity. One illustration is the IBM project to create a business-data network with direct access to a communications satellite by many participants.

According to Nora and Minc, the French government is confronted with the task of seeking a delicate equilibrium between a positive balance of payments and full employment. For example, the authors estimate in their report that the new information technology will enable the French banking and insurance industries, employing 600,000 workers in the late 1970's, to reduce their workforce by 30 percent by 1990. Nora and Minc suggest that the private sector will be unable to create the necessary increase in employment to compensate for this loss in view of growth in productivity. Hence the government may be obligated to provide more jobs in the public sector. Ultimately, these are financed by taxes on the private sector, which is subsequently weakened, leading to a deterioration of the balance of payments. It is essential for the government to strike the right balance between unemployment and deterioration in the balance of payments. The resulting question is whether a temporary decline in the standard of living is acceptable to avoid social unrest caused by a high rate of unemployment.

If telematics is a key factor in the effort to improve productivity, as Nora and Minc postulate, this technology should be made available to interested parties in France on a nondiscriminatory basis. The implication is that the essential communications networks cannot be left to a few commercial interests, since this would result in an unacceptable private influence, for instance through tariffs or compatibility requirements for certain types of equipment.

Thus, as a consequence of its responsibility to strive for maximum employment, the French government is urged to support the rapid development of telematics. An equilibrium of "power and counter powers" must be sought to prevent a one-sided access to the new technology. The nation's freedom to adopt such policies, however, is limited as a result of the virtual monopoly of IBM in the computer market and its increasing command of digital networks. However, European countries exercise a strong influence over their telecommunications network through their PTT organizations. Nora and Minc believe that this advantage should be strengthened by extending the PTT activities to modern communication technologies. This step would entail launching and exploiting communications satellites and managing data banks. This resulting improvement in the bargaining position of European countries vis-a-vis IBM is deemed essential for solutions which serve their national interests.

For France, such a strategy requires a coordinated and highly focused telecommunications policy. An industrial policy is needed to strengthen France's abilities in the telematics field: for instance, small computer makers, software industry, and terminal manufacturers should be supported. Research and development should be devoted to computer science and integrated circuits. The state's own use of telematics should be promoted, and strategies must be devised to ensure that this will lead to a further democratization of society.

Barron, in his study entitled "The Future with Microelectronics,"(5) also concludes that the most immediate consequence of the technology will be its impact on employment, particularly in the service sector. In the past, in general only a small amount of capital was invested in increasing the productivity of office workers such as secretaries, typists, clerks, and managers. It is expected that the application of electronic information technology will have a marked effect on the level of productivity in the office, and therefore on patterns of employment. According to Barron, the widespread introduction of the so-called "office of the future" (see also Chapter 3) will have substantial labor-displacing effects. The trend toward distributed data processing, and to direct access to data in data banks, will have an adverse impact on clerical occupations and "reduce the middle or low-level managerial skills used to initiate or monitor such activities." Barron observes that 65 percent of the working population is employed in the "information occupations." Thus, "even moderate improvements in productivity could bring about unemployment levels in the 10 to 20% region unless offset by compensatory increases in demand in these other activities."

What Barron and other researchers who have examined this problem find particularly disconcerting is the present European economic framework into which microelectronics is being introduced. Labor displacement will take place in a society already experiencing slow growth. Both white-collar and blue-collar workers will be affected in Barron's scenario. He fears that the "microelectronics revolution" will have a cumulative effect on employment which may be recognized too late, precluding any effective public-policy solutions.

Barron proposes that a public policy be created concerning microelectronics which would include educational programs. Included here would be vocational training in information technology and adult re-education programs for workers affected by the introduction of microelectronics technology.

Many other studies on the impact of microelectronics upon employment have been undertaken by industrialized countries as well as by OECD. It is generally forecast in these studies that substantial numbers of the existing labor force will lose their jobs as a result of labor displacement. Additionally the content and/or the quality of many jobs will change. Prentis notes that "The occurrence of technological unemployment on the scale predicted by some analysts would clearly be unprecedented in the world economic history of at least the last two centuries."(6) This author concludes, however, that the fear of widespread unemployment resulting from the application of microelectronics technology is exaggerated.

One OECD study, in discussing the potential impact of microelectronics, concludes that the development will be evolutionary rather than revolutionary and therefore should not cause disruptive social change.(7) This report proposes that industry, management, labor, and government should join in consultation during a transition period leading to a new economic infrastructure. Some areas, such as telecommunications and office automation, will experience the most rapid changes. There will be a constant need for retraining of manual and office workers. The report states that public education, both for school children and adults, will be needed to accommodate the technological changes.

According to this OECD report, it should not be assumed that there will only be adverse employment effects as a consequence of microelectronics. Rather, the authors of the study believe that a good case can be made for a scenario showing positive employment effects in the medium to long term. They state that both in the U.S. and Europe high-technology industries have in the past created more new employment than low-technology industries. A study is cited, conducted by Arthur D. Little, which projects one million new jobs in the electronics industry in the years 1977-87. Citing this study, the OECD experts believe that new industry will create several hundred thousand jobs in Europe.

The problem of distinguishing between employment effects of technological change and unemployment which is directly related to current economic conditions is discussed by Smith.(8) Studies which quantify and separate the various components of the unemployment problem are needed. In the absence of such studies, debate is usually focused on technological unemployment as the major problem facing governments in the future.

However, technological displacement is generally a relatively small portion of normal frictional unemployment. The traditional view is that technological displacements would be more than compensated for by the stimulative effect on aggregate demand of accelerated technological progress.

Smith notes that the relationship between technological change and overall employment is a complex one. One commonly held view is that technical change permits greater worker productivity. Consequently, the number of workers required declines, and therefore unemployment must rise. This view assumes that the level of output remains constant after technical change occurs. Smith believes that this is often not the case. As a result of increased productivity, "real costs of output fall and real incomes rise and these changes lead to increases in demand and output," for both the particular product affected as well as for all goods.

Technological change does, however, result in the reallocation of two factors of production: labor and capital. Particular occupations, industries, and regions will be affected and displaced workers will be forced to make adjustments. This adjustment process will occur even at times of high levels of economic activity and low levels of unemployment. Smith thinks that the combination of technological change and the present relatively low levels of economic growth, rather than the technology per se, results in unemployment following the introduction of the new technology.

Certain groups such as women and semiskilled workers will be affected more than others. Women, who are disproportionately concentrated in occupations where the application of microelectronics will probably be the most rapid, will suffer more than men in terms of job displacement. Young people and unskilled and semiskilled workers would also be particularly affected.

However, despite all of the potential impacts of microelectronics in the workplace, Smith notes that its application will not be without limits. What is technically feasible is not necessarily economically viable or socially acceptable. Thus, though the rate of diffusion will vary among occupations, most changes will be evolutionary, spanning the next two decades.

Summarizing, Smith states: "Despite widespread claims that microelectronics will have catastrophic effects on aggregate employment there is little evidence to support this view. There are estimates of the extent of job displacement which is likely to occur. Case studies of individual industries, however, give no guide to the aggregate employment effects of technological change. They tend to focus on job destruction in the industries under review and ignore the possibility of job creation elsewhere."

Other observers believe that the developments in microelectronics present a number of unique characteristics which have virtually no precedent in the history of technology. For example, Kendrick(9) notes that there may be a major difference between the adoption of earlier technologies and the speed for microelectronic product and process innovations. The speed of their diffusion and the breadth of their applications are not comparable with what happened during earlier major technological developments. It would seem, therefore, "wise to devote considerable resources to a further study of the developments and preparation of policy options to mitigate the probably undesirable side-effects that lie ahead."

In assessing the impact of microelectronics, there are various constraints on the rate of growth of production of microelectronic devices and on the speed of diffusion of the various technologies that incorporate them. For instance, it takes time to develop managerial capabilities, to educate and train personnel, and to build plants and equipment to produce the intermediate and final goods and services. More broadly, there are limitations on total saving and investment and on the ability of any particular sector to finance a new technology. These factors will limit the rate at which microelectronics is applied.

Since not all labor displacements can be accommodated within firms and thus layoffs are inevitable, Kendrick suggests that government programs for retraining be improved. It is also important that curricula of universities and technical schools be restructured so that students are educated and trained for the kinds of positions that will be available. In the long run the more basic problem may be that too few jobs will remain for unqualified workers.

In the United States, scholars of automation do not expect a rapid substitution of machinery for human labor to increase unemployment, assuming healthy economic growth. But they do expect a radical restructuring of work, including a devaluation of current work skills and the creation of new ones at an ever-increasing rate. Ultimately, the nation's educational system will have to prepare workers for functioning in an electronic society. More quantitative studies will have to be conducted on the effect of microelectronics on total employment and on the spectrum of occupations that will become available. In this respect Freeman has proposed that empirical work in every sector of application and development be carried out in order to assess the probable employment consequences over the next ten years.(10) This program has already begun on a small scale at the Science Policy Research Unit at the University of Sussex, England.

Robinson(11) sees the future labor force as being composed of large numbers of relatively unskilled workers at one end and highly trained managers and engineers at the other, with very few medium-skilled individuals in between. Robinson points to the example of the installation of electronic switching systems at Western Electric as an illustration of this point. However, other models of resulting job distributions have been suggested.(2)

GOVERNMENTS AND LABOR UNIONS

Most European governments, recognizing the wide range of consequences of the "microelectronics revolution" discussed above, have commissioned policy studies. Typical examples are a British study(12) of the Advisory Council for Applied Research and Development and the Dutch study by Rathenau.(13) A study more directly concerned with employment is the one by Sleigh.(14) The proposed measures fall generally into the following categories:(15)

- Reduction of work time or sharing of jobs.
- Stimulation of new industries and products and setting up R&D programs.
- Establishment of incentives for small and medium-sized industries, which are expected to provide new jobs (since the bigger organizations are more interested in rationalization of investments).
- Creation of employment in the service sector by expanding the civil service.

Positions regarding such measures have been taken by employers(16) and labor unions.(17,18) The former usually stress the opportunities that are available and suggest that proper policies be established by government to enable industry to successfully adopt new technologies. The labor unions usually require consultation when management wants to introduce new technology. To this end a formal "New Technology Agreement" has been proposed, supplementing the usual labor-contract agreements. In their policies concerning labor productivity, labor unions in Western Europe will consistently seek a shortening of the work week as a measure against loss of jobs.

Western European governments are required to map out their course regarding the unemployment issues. As we pointed out before, this requirement can lead to a very broad range of policies including industry, education, trade, and industrial relations.

However, short-term problems often tend to affect the implementation of long-term strategies. In particular, such short-term problems will arise in the job market in a period of rapid technological change. Though it can be argued that net effects over several years will be favorable, the temporary occurrence of unemployment in particular sectors, or in certain categories of job qualifications, may force governments to respond with actions that have immediate results. Such actions may prove detrimental to the government's own strategic socioeconomic goals.

Finally, relations between countries are influenced by the emergence of microelectronic technology. The present competition between the U.S., Japan, and Europe is receiving most of the attention. However, relations with newly developing countries such as Brazil and South Korea as well as with the underdeveloped nations may increasingly lead to concern.(15,19)

REFERENCES

1. Rothwell, R., and Zegveld, W., Technical Change and Employment, Response prepared for the six countries programme on government policies toward technological innovation in industry, 1979.

2. Bundesministerium fuer Forschung und Technik, Information, Technologie und Beschaeftigung (Econ Verlag, 1980).

3. Nora, S., and Minc, A., L'Informatisation de la Société, (Paris: La documentation Française, 1978).

4. Bell, D., "The Social Framework of the Information Society," in Forrester, T. (ed.), The Microelectronics Revolution, (Oxford: Basil Blackwell, 1980).

5. Barron, I., The Future with Microelectronics, (New York: Nichols Publishing Co., 1979).

6. Prentis, M., "The Impact of Information Technology on Employment in Canada: A Review of Current Research," in Microelectronics, Productivity and Employment (Paris: OECD, 1981).

7. Group of Experts on Technology, Business and Industry Advisory Committee, "Impact of Microelectronics on Employment," In Microelectronics, Productivity and Employment" (Paris: OECD, 1981).

8. Smith, J.S., "Implications on Developments in Microelectronic Technology on Women in the Paid Workforce," in Microelectronics, Productivity and Employment (Paris: OECD, 1981).

9. Kendrick, J., "Analytical Summary and Perspectives on the Impact of Microelectronics on Productivity and Employment," In Microelectronics, Productivity and Employment (Paris: OECD, 1981).

10. Freeman, C., "Unemployment and Government," in The Microelectronics Revolution, Forrester, T. (ed.) (Oxford: Basil Blackwell, 1980).

11. Robinson, A., "Electronics and Employment: Displacement Effects," in The Microelectronics Revolution, Forrester, T., (ed.), (Oxford: Basic Blackwell, 1980).

12. Advisory Council for Applied Research and Development, Technological Change, Threats and Opportunities for the United Kingdom, (London: HMSO, 1979).

13. Rathenau, G., et. al., Report of the Advisory Group on Microelectronics, Dutch Government Printing Office, 1980.

14. Sleigh, J., et al., The Manpower Implications of Microelectronic Technology, (London: HMSO, 1979).

15. Rada, J., The Impact of Microelectronics (Geneva: International Labor Office, 1980).

16. Confederation of British Industry, Jobs Facing the Future, (London: 1980).

17. Association of Professional Executive, Clerical and Computer Staff, Office Technology, The Trade Union Response, London, 1979.

18. Trades Union Congress, Employment and Technology (Interim Report) (London, 1979).

19. Norman, C., Microelectronics at Work: Productivity and Jobs in the World Economy, Worldwatch Paper 39 (Washington, D.C.: 1980).

20. "Japan's Strategy in the 80's: A Changing Workforce Poses Challenges," Business Week, Dec. 14, 1981, p. 166.

21. "Research and Study Concerning Influences of Microelectronics on Employment," Ministry of Labor. Published by Foreign Press Center, Japan (R-81-03), 1981.

11
Policies of Federal and State Governments

The U.S. federal government is in many ways confronted with microelectronics and its effects. Let us review some of the major issues.

First, a leading capability in microelectronics is essential for the U.S. This is obvious regarding national security, which in many ways relies on superior electronics technology. But it has also become increasingly clear that broad sectors of industrial activity will rely on microelectronics for their future competitiveness, be it regarding either the sophistication of products or the productivity increase that it allows.

Thus the federal government is confronted with options to support a strong microelectronics capability. Several policies are available to this end:

- Industry may be supported by creating a healthy technology base, either through the government's R&D facilities, through stimulating academic programs, through R&D contracts or purchasing policies, or by removing certain regulatory barriers that prohibit R&D cooperation between companies.
- Alternatively, industry might be protected from competition by foreign companies, for instance through import quotas or special duties. Another option is to restrict the flow of high-technology know-how and expertise by classifying information or applying export-control regulations, designed to safeguard a perceived U.S. lead in technology.
- The federal government might also promote free-trade policies internationally, with minimal government interference. This policy would entail limiting foreign government support of indigenous industries, achieving adequate access to government-controlled markets abroad, and so forth.

Policies like these have to be carefully adjusted in order to achieve the desired effects. This is not always easy in view of the complexity of the issues at hand. A few examples:

- The merchant industry's competitive position is internationally strong in some product areas but is threatened by competition in others. Evidently competition is a healthy phenomenon, so we must ask, "Where and why is it really threatening the vitality of the industry?" And when this is defined, the measures to be taken must be carefully analyzed. Would protective measures be effective in such an international business, certainly with foreign ownership of a substantial number of manufacturing facilities in the U.S.? Is restricting the flow of technology a good idea, with part of the industry in the hands of multinational companies which rely heavily on the flow of technical information for their activities? Also, would it frustrate the traditional mobility of scientists and engineers that has had such a favorable effect on the growth on the industry?
- Technical change in IC technology is still substantial, and it continues to have important effects on industry growth and industry structure. Increasingly, systems aspects take precedence over the components per se. In terms of dollars, the systems industry is much more important than the component industry. The former has been extensively engaged in captive IC manufacturing for its special needs and buys on the world markets whatever it wants in addition. Measures of the federal government designed to support the component industry may be detrimental to the U.S. systems houses, a situation that may become more pronounced as the foreign presence in the markets grows. A related issue is that in the future the importance of a leading technology will become less pronounced; much more emphasis will be given to capabilities in designing hardware and software. Thus, technologically oriented support programs may be outdated soon if this trend continues.

Another class of policies is related to the application of microelectronics. A wide variety of issues is at stake, some of which have already received some attention.

- The potential of connected computers with access to data banks raises questions as to the protection of individual privacy. Legislation in this field has been drafted or proposed in most industrialized countries. With further use and availability of data networks, extending to the home environment, such regulations will have to become

stricter. Also, legislation dealing with computer fraud and related crimes will need adaptation.
- Monopolization of markets like that of telecommunications must be avoided. In most European countries governments exert a decisive influence on these markets through their PTT's, while in the U.S. the Federal Communications Commission provides the necessary regulations. Prompt adaptation to newly evolving communication systems will be necessary.
- Changes in employment will be brought about by microelectronics, as we discussed in some length in the preceding chapter. Since hard facts are lacking, there is ample opportunity for positions favoring either complete laissez faire or a strong government control of introduction of new technology.

Also in these issues, difficult problems occur which are closely related to aspects of the discussions in earlier chapters. For instance, would measures against monopolization weaken the competitive international position of the U.S. companies involved? How will services like electronic banking, in particular home banking, be developed, and when will fiber-optical communication networks become available? How should education, training, and retraining needs be organized in the wake of technological change: do private and state initiatives provide sufficient momentum?

In summary, microelectronics technology gives rise to an endless list of potential issues for federal government attention where regulation or legislation might be required. Just how many tasks the federal government is willing to assume depends strongly on philosophy and tradition. But it is evident that factors like technology change, industry growth, and social impact enter into any policymaking process.

A proper understanding of potential benefits and threats is essential to this discussion. Notions of the "microchip" as either a knight heralding new prosperity or as a villain responsible for the destruction of our traditional values are obviously oversimplifications. At stake is how microelectronics can be put to optimal use, in the meantime with the reduction of unfavorable side effects.

ATTITUDES OF THE FEDERAL GOVERNMENT

Let us now examine several federal policies that affect the microelectronics industry. The attitude of the federal government toward business in general has been subject to considerable change since World War II. As Herzstein[1] has pointed out, the development of international trade policies has been a major determinant.

Strongly stimulated by the U.S., international measures were taken to liberalize world trade. A major milestone was reached in 1948 with the General Agreement on Tariffs and Trade (GATT). This agreement liberalized world trade by instituting a systematic tariff reduction and a code of conduct for governments regarding international trade.

In 1949 the U.S. restricted trade in strategic goods and general exports to Communist countries with the passage of the Export Control Act. However, the drive toward furthering free trade in general was supported by the federal government. This resulted in the Kennedy Round (1962-67) and the Tokyo Round (1974-79). International agreement was obtained on a continued reduction of tariffs and on codes that reduce nontariff distortions on trade, such as subsidies, countervailing duties, standards, and government procurement.

This international development has resulted in a trend toward removing governments from international business. This has been the case also in the United States. The federal government, in principle, pursues a noninterference policy regarding the microelectronics industry. In practice, however, many federal policies such as those on regulation, export limitations of strategic materials, and procurement for national security have strongly affected the industry.

In recent years the Japanese semiconductor industry has posed a serious threat to U.S. manufacturers. Strongly influenced by coordination and subsidies, directed by the Ministry of International Trade and Industry (MITI), they have developed a highly competitive technology and have captured an important share of the international markets. This threatens the profitability of the U.S. manufacturers who need to be able to invest sufficient capital for their future operations.

In light of the general trade principles discussed earlier, debate on the Japanese threat and the federal government's response is concentrated on the following issues: (1,2)

- The Japanese import duty on semiconductors has been much higher than that internationally agreed upon, although a decrease of this rate was announced in 1981.
- The Japanese industry seems to have several distinct advantages such as government support for R&D, easy terms on loans, tax deductions for investment, higher debt-equity ratios, and the ability to operate with much lower earnings.
- The Japanese have been accused of using such business methods as price dumping and targeting practices.
- The Japanese government has strictly limited American access to Japan. A "buy Japanese" policy is pursued by such government agencies as Nippon Telephone and Telegraph (NTT) and is encouraged as a national policy.

Of course, these issues are not all clear-cut. American industry often assumes that the American way of doing business should be the world standard. Many of the Japanese practices are the result of the economic structure and industrial culture of Japan.(3) For instance, close ties between government and industry or industry and banking are normal and acceptable in Japan while in the U.S. the situation is vastly different. Both systems have their advantages and disadvantages and both must function within their indigenous cultures.

Trade relations between the U.S. and Japan have been investigated by the so-called Wise Men Committee consisting of U.S. and Japanese representatives. Comparative studies of Japanese and American industrial practices are being conducted at Stanford(4) and Berkeley.(5) They may lead to a better understanding of questions related to fair competition between the two countries.

The federal government has moved with caution regarding support of the IC industry. This holds for the trade-oriented policies of the Department of Commerce and State Department and the more technological role of the Department of Defense and the National Science Foundation.

DEPARTMENT OF COMMERCE

The stated policy of the Department of Commerce is to promote the healthy development of American industry by improving its competitive position in international markets. In this respect, it tries to foster the free-trade procedures that were agreed upon internationally. Thus the Department will take action with antidumping measures and countervailing duty laws only when it is viewed as necessary to keep foreign nations from violating the trade rules.

The way in which such general policies are implemented depends strongly on the political outlook of the incumbent administration. Under President Carter the Department of Commerce moved toward a more systematic industrial-sector policy. For instance, analyses were prepared of the semiconductor industry to identify its strengths and weaknesses in preparation for the development of such a policy for the industry. An effort was initiated to get better statistical data on international trade in semiconductors. A program for the common industrial development of generic technologies, COGENT,(6) was established and considered for application in the semiconductor industry.

However, such actions by the federal government were not generally applauded by industry. Several industry spokesmen wish to restrict government action to the areas of tax cuts for capital investment or R&D expenditures, and to policies which can open the Japanese market for U.S. industry.

Under the Reagan administration the trend toward more government influence on industry has been reversed. Business will be left alone to a greater extent and supported mostly by across-the-board tax cuts for capital investment and possibly for R&D. The sectoral analyses and the COGENT program have been phased out. In the near future there does not seem to be much promise for any special measures that would specifically benefit the semiconductor industry. The "fairness" principle may be expected to determine foreign-trade policies.

Trade is affected more specifically by the Department of Commerce in implementing its Export Administration Regulations. These control U.S. technology exports not otherwise controlled by the Atomic Energy or Arms Export Control Acts. These regulations also cover not only high-technology components and equipment but the export of technical data as well.

The National Bureau of Standards(7) falls under the responsibility of the Department of Commerce. In addition to its program on semiconductor technology which we discussed earlier, the NBS is stressing its role in the nation's efforts toward increasing productivity and innovation. In view of the crucial role of microelectronics in both areas, a sustained high level of attention for this field seems likely.

STATE DEPARTMENT AND JUSTICE DEPARTMENT

The State Department is instrumental in implementing trade policies abroad and also facilitates international agreements on scientific and technological exchange programs. The science attaches in major embassies provide a continuing review of scientific and technical developments abroad for information to the government.

An important area of concern for the State Department has been the limited access of U.S. firms to the Japanese market. In this respect the Department, through the embassy in Tokyo, has been active in the following areas:(1)

- Changing the high import tariffs in Japan.
- Overseeing customs valuation and procedures.
- Improving U.S. access to Japanese technology.
- Discussing Japanese limitations on foreign investment.
- Liberalizing government procurement policies, especially those of Nippon Telephone and Telegraph.

FEDERAL AND STATE POLICIES

A number of these actions have resulted from pressure by the semiconductor industry. However, the State Department tries to avoid the use of international trade policies to address domestic problems which originate in structural differences between countries.

The State Department is also responsible for the International Traffic in Arms Regulations(24) (ITAR).

The Justice Department enforces antitrust laws, which, for instance, limit possibilities for companies to coordinate or share industrial research activities. A major case against AT&T led to the well-known consent decree of 1956, which promoted the diffusion of transistor technology, an invention of Bell Laboratories.(14)

More recently, proposals to deregulate parts of the AT&T business,(19) initiated in Congress and by the Federal Communications Commission, have important antitrust aspects. Another example is a lawsuit against IBM, which was accused of monopolizing the computer market.

Early in 1982, the AT&T case was settled by an agreement on a divestiture of its local telephone companies in exchange for access to the deregulated communications business. At the same time the antitrust case against IBM was dropped. These decisions will probably have a major influence on the future U.S. telecommunication and computer markets, which are gradually merging in several areas.(26) Strong competition between these giant companies is expected to evolve, AT&T entering these new markets from a telecommunications background and IBM coming from computers. Abroad, AT&T will probably become a more aggressive IBM-like international competitor.

The position of the Justice Department regarding cooperative R&D activities among companies is not yet clear. The IC business established a Semiconductor Research Cooperative through which research at universities and non-profit centers will be sponsored. Several computer and IC companies, led by Control Data Corporation, have discussed the possibility of founding a company that would carry out advanced product development.(25)

NATIONAL SCIENCE FOUNDATION

NSF, since its establishment in 1950, has strongly focused on the pursuit of basic science at universities. Though other government agencies are also major providers of research grants, NSF is considered by universities as "their" institution. This orientation is reflected in its management (program directors are often recruited from university staff and some arrange to return after a limited time at NSF), and in the funding procedures (grants are awarded to unsolicited and solicited proposals, judged by a peer-review system).

In the last decade an evaluation of NSF's contribution to American society has received increased attention. This was stimulated by the prevailing mood of the general public, which questioned the necessity of big spending on basic science projects. The need to defend NSF activities against such pressure from Congress has led to a stronger interest in research programs that might have a more applied benefit in economic terms.

Several programs were established aimed at using basic research knowledge in solving problems in society. Programs in solar energy and earthquake prediction fall into this category. Later, other mission-oriented programs directed at national needs were begun. However, such efforts often interfered with programs of other agencies.

After some reorganization, NSF established a Division of Engineering Sciences, mainly designed to support the needs of engineering schools. These schools had been somewhat neglected by the prevailing NSF policies,(23) and the new reorganization may correct this imbalance.

Not surprisingly with NSF's "academic" orientation, it has traditionally had few ties with industry. Several programs have been established to improve the relation between scientific and industrial research, but the level of such efforts has been rather modest.

In industrial circles some criticism can be heard about NSF's way of operation, which merits a brief discussion. Some arguments pro and con NSF's modus operandi are:

- NSF management is recruited from the academic world, as seems reasonable with respect to the agency's mission. However, this leaves little interest in and sensitivity to industrial problems.
- The peer-review system is thought to promote too much self-interest. (Only 10 to 15 percent of the reviewers are affiliated with industrial laboratories). However, another judgment system in pure science is difficult to design. Research has indicated that rather than a reviewer's bias, chance is a major factor in getting a research grant.(20)
- No systematic feedback of results exists. Though this is factually true, scientific quality is usually reflected in the estimate of peers for one's work. Also, universities look strongly at quality when considering tenure.
- The NSF programs are incoherent; a "master plan" does not exist. On the other hand, the direction into which science develops is strongly determined by the judgments of informed individuals.

FEDERAL AND STATE POLICIES

The Reagan administration is effectively reducing NSF's budget. This holds also for several other government agencies sponsoring academic research, with the exception of the Defense Department. Industrial sponsors in some cases replace research funding by government. As Gray(8) notes, this implies a return to university-industry relations which were typical of the period during and shortly after World War II, before government became the major supporter of academic research.

DEPARTMENT OF DEFENSE

The Department of Defense has been a major factor in the development of integrated circuits, as Utterback and Murray have documented.(9) They have shown that the direct and indirect procurement of IC's for the Minuteman II missile and Apollo project provided a firm basis for several companies to develop necessary new technologies. These efforts stimulated a large amount of company-sponsored R&D, greatly surpassing the size and influence of the industrial R&D directly sponsored by Defense.

In addition, military-program directors in those days were often given credit for their flexible attitude, which allowed an optimal utilization of industry's new development. In later years, more bureaucratic rules were imposed with an ensuing loss of flexibility.

In recent years, Defense has followed the development of the commercial IC market, deriving militarized IC versions where needed.(9) During this process the fraction of military applications in the total market has dropped sharply to approximately 7 percent in 1980. (However, military IC market is expected to gain importance in the 1980's in view of the increased U.S. defense budgets.)(21) This decrease led to some reluctance in industry to devote resources to this market since they could be applied much more profitably for the civilian market. Therefore, Defense established the Very High Speed Integrated Circuit (VHSIC) program, aimed at the development of very advanced programmable IC's for military-signal processing systems.(10)

The Defense Department's attitude toward basic research has changed in recent times after a period of continuous reductions in defense spending for university research.(11) The reduced funding levels generated concern that the future use of basic university research for military applications would be impaired. Funding has since been increased, a special coordinating director nominated, program reviews initiated, and red tape reduced.(12) A further increase in Defense-sponsored research can be expected in view of the commitments of the Reagan administration.

The growth of defense research has posed some direct problems to the academic world regarding security requirements.(27) Of course the broader question remains whether a university research program should be directed by national defense requirements. It is similar to the question of university involvement with industry, although with a different emotional content and different implications. Ideally, each university should be able to follow its own course in these matters, which implies that a certain diversity in potential funding sources is desirable.

The imposed security measures of the Department of Defense and the Department of Commerce have generally had only a modest impact on industrial R&D, since company policies have generally been sufficient to protect proprietary information and safeguard against the undue transfer of technology. However, more stringent security measures are being considered which would control not only the export of tangible products but also exchanges of technical data.(13,24,28) These could adversely affect the development of new technical advances and the vigor of international competitiveness, which could in turn adversely affect national-security objectives.(24) Earlier studies(14,15) of how technology transfer occurred in the semiconductor industry have illustrated the essential role of a free flow of technology and its relation to technical and commercial advances.

THE SIA AND PUBLIC POLICY

The Semiconductor Industry Association (SIA), established in 1977, unites most major merchant semiconductor houses. (The industry leader, Texas Instruments, is not a member.) Several captive producers like IBM and Honeywell are also members. It has engaged in activities to promote its viewpoints in the administration and Congress, and to coordinate the relations between industry and the academic world.

The SIA, as an organization, began its work by organizing a system of trade statistics in which its membership participates on a voluntary basis. Soon thereafter, the association became active in formulating and promoting industry's view of the task of the federal government, mainly in relation to Japanese competition. It has organized a number of discussions between representatives from government and industry, aimed at obtaining a better mutual understanding.

In a recent position paper,(2) the SIA identified the following public-policy issues.

Market Access

- Reduction of the Japanese import tax to the U.S. level and a subsequent bilateral elimination.
- Equal national treatment, i.e., the acceptance of U.S. ventures in Japan, access to MITI-sponsored R&D, and open purchasing policies.
- Reduction and subsequent elimination of the current 17 percent EEC import duty.
- Liberalization of rules of origin in Europe.
- Elimination of restrictions on government purchasing.

Capital Formation

- A reform of depreciation methods is proposed, which takes into account the rapid technological obsolescence.
- Incentives for all R&D should be encouraged, including measures to promote industry support of university research.

Research and the Development of Human Resources

- Mobilization of universities to expand their education of engineers and scientists, and to induce universities to do more research in supporting the semiconductor industry. This could be accomplished either by tax incentives for industrial grants to universities or by direct government grants.
- Establishing industry-university steering committees to advise on the direction of university research.

SIA continues to work along these lines. The association attempts to formulate the industry's viewpoint in these matters and to disseminate them in legislative and executive branches of the government. For instance, it vigorously promotes those tax incentives which are favorable for high-technology, high-growth industries. (In 1981 a new tax law was passed providing a 25 percent credit for certain incremental R&D activities. The effects of this new law on high-technology companies are still far from clear.)(22) As Knickerbocker(2) notes, priorities of this sector are different from those of mature industries like the automotive and steel industries. These industries are more interested in tax reductions that produce a direct cash flow than in tax incentives for R&D. Moreover, the mature industry has considerable political power, in part because of strong interests of labor unions. Because the new industrial jobs have generally been created by the younger, high-technology industries, SIA feels that public policy should pay more attention to their needs.

Perceiving insufficiencies in the academic contributions to the technology base, SIA also initiated an investigation of the possibility of a cooperative approach to the university research problem, resulting in the establishment of the Semiconductor Research Cooperative, which will sponsor certain types of university research.

POLICIES OF THE STATES

Several states are attempting to attract microelectronics industries, since they provide high-quality work with little burden on the environment.(17) These efforts can be facilitated by congestion problems (e.g., housing costs) in the traditional areas of the industry such as California and Massachusetts.

A number of IC manufacturers have established new facilities in states such as Colorado, Arizona, and New Mexico. North Carolina has played a notably active role in attracting industry through the establishment of the Research Triangle Park and its support for university research. Several other states, like Ohio and Minnesota, are also operating along similar lines.

In 1981, Governor Edmund G. Brown, Jr., of California proposed a series of measures to counteract this lure from other states.(16,18) These would include a strengthening of the University of California at Berkeley, subsidies for innovative private research, support for new ventures, and assistance in industrial reinvestment programs.

CONCLUDING REMARKS

How will future federal policies respond to the continuing technical change brought about by microelectronics? We expect that the general trend will be toward more involvement of the U.S. government, since it has to respond to the following developments:

- Microelectronics is assuming the role of a key technology in national defense, as a technological basis of large sectors of industry, and as a means to enhance productivity. A strong U.S. position among other postindustrial societies hinges on its microelectronics capability.
- Global competition is emerging strongly in microelectronics. Many foreign governments have great influence on the development and application of the technology

through such activities as R&D grants, industry subsidies, trade policies and directed purchasing. Though perhaps deviating from its traditional stand, the U.S. government may feel obliged to resort to similar measures.
* Microelectronics is very quickly finding broad application in society. Novel technologies create new situations in which existing regulation and legislation are not adequate. Since many people are likely to be affected, new initiatives in this field must be taken by the federal government.

It is beyond the scope of this book to speculate on the direction of novel policies. At any rate, they will often have to reconcile seemingly opposing viewpoints. For instance, support for industry would conflict with the U.S. tradition of open trade. Allowing companies to share R&D may seem inconsistent with antitrust policies. Regulation of the introduction or use of microelectronics would not fit in an economy so strongly based on free enterprise. Very complex but very interesting issues await the federal government during the coming decade. Several questions which already merit further consideration and study will be explored in the following chapter.

REFERENCES

1. An American Response to the Foreign Industrial Challenge in High Technology Industries. Conference at Monterey, CA, June 1980. (Cupertino, CA, SIA, 1980).

2. Semiconductor Industry Association, The International Microelectronic Challenge. The American Response by the Industry, the Universities and the Government, (Cupertino, CA: SIA, 1981).

3. Gresser, J., Statement Before the Committee on Ways and Means, U.S. House of Representatives, Subcommittee on Trade: "High Technology and Japanese Industrial Policy: A Strategy for U.S. Policymakers" (Washington, D.C.: 1980).

4. Project on U.S.-Japan Relations, Northeast Asia-U.S. Forum on International Policy, Stanford University, to be published.

5. Zysman, John, to be titled and published. University of California, Berkeley.

6. COGENT, Department of Commerce Fact Sheet, Washington, D.C., Oct. 1, 1981.

7. Charles River Associates, "Productivity Impact of R and D Laboratories: The N.B.S. Semiconductor Technology Program," (Boston: CRA, 1981).

8. Gray, P.E., "MIT Wants Closer Ties with Business," The New York Times, Sept. 27, 1981.

9. Utterback, J.M., and Murray, A.E., "The Influence of Defense Procurement and Sponsorship of Research and Development on the Development of the Civilian Electronics Industry." Center for Policy Alternatives, MIT, CPA-77-5 (1977).

10. Davis, R.M., "The DoD Initiative in Integrated Circuits," Computer, July 1979, p. 74.

11. Lepkowski, W., "Defense Department Boosts Research Funding," Chemical and Engineering News, April 27, 1981, p. 14.

12. DoD Basic Research Program. DoD publication, August 1980.

13. Lombardo, T.G., "Technology; Dichotomous Tool." IEEE Spectrum, May 1981, p. 51.

14. Tilton, J.E., International Diffusion of Technology, the Case of Semiconductors, (Washington, D.C.: Brookings Institution, 1971).

15. Finan, W.F., The International Transfer of Semiconductor Technology Through U.S.-Based Firms. Preliminary Report, National Bureau of Economic Research, Washington, D.C., 1975.

16. Science, "Governor Brown Boosts Microelectronics," Feb. 13, 1981, p. 688.

17. McInnis, D., "Going After High Tech," The New York Times, July 19, 1981, p. F6.

18. Lueck, T., "Governor Brown Seeks Business Aid," The New York Times, February 12, 1981.

19. Uttal, B., "How to Deregulate AT&T," Fortune, Nov. 30, 1981.

20. Cole, S., et al., "Chance and Consensus in Peer Review," Science 214 (1981), p. 881.

21. Stahr, L.B., "Military IC Market To Win Big From Increased Defense Spending", Defense Electronics, August 1981, p. 61.

22. Feder, B.J., "The Research Aid In New Tax Law", The New York Times, Sept. 29, 1981.

23. Weinschel, B.O., "Proposal: A National Engineering Foundation," IEEE Spectrum, February 1980, p. 58.

24. Haklisch, C.S., and Fusfeld, H.I., "Current Issues in Export Controls of Technology," Center for Science and Technology Policy, New York University, November 1981.

25. "A For-Profit Lab to Help Chip Makers Compete," Business Week, April 5, 1982, p. 35.

26. "The Odds in a Bell-IBM Bout," Business Week, Jan. 25, 1982, p. 22.

27. Gray, P.E., "Technology Transfer at Issue: The Academic Viewpoint." IEEE Spectrum, May 1982, p. 64.

28. Wallich, P., "Technology Transfer at Issue: The Industry Viewpoint," IEEE Spectrum, May 1982, p. 69.

12
U.S. Microelectronics

In this book we have described microelectronics as a technology-driven phenomenon with far-reaching economic and social consequences. This technical change has been and will continue to be based on the progress achieved in integrated circuits. As discussed in earlier chapters, simultaneous cost reduction and performance increase of IC's have caused an unprecedented technical revolution.

It may be pointed out in review that a key concept has been the advance in digital techniques. These emerged in electronics from research on nuclear physics but found their first widespread use in computers. Later on, digital techniques found applications in electronics handling analog signals: a digital representation of such signals can be transmitted, processed, and stored with excellent accuracy.

These developments have resulted in a large common digital technology base which is usable in many areas of electronics. Digital IC's form the building blocks for these techniques. One type can often be applied for a wide variety of data-handling and signal-processing tasks. This multiple applicability of many digital IC's results in the possibility of mass manufacturing and, thanks to the learning-curve effect, low-cost production.

This availability of a wide range of low-cost digital IC's has opened possibilities like the development of novel equipment, electronification of mechanical functions, and increasing the capabilities of existing electronic equipment. Microelectronics-based equipment has become very attractive for use in the home, office, factory, and many other places. The reactions to this process are manyfold and, as discussed in the preceding chapters, are the result of a complex interaction involving many concerned parties.

One particular factor, however, may not have received sufficient attention: the public acceptance of this new microelectronics technology. Here we are not concerned solely with the attractiveness or usefulness of a particular piece of electronic equipment or system. Neither are social and political questions about productivity and employment primarily relevant. Rather, the key question is whether the equipment or system fits in or implies changes in an individual's life or an organization's modus operandi. Little is understood about how a person's rational and irrational behavior may be related to the acceptance of novel concepts in electronic aids, or to how this changes from generation to generation.

In this connection, some caution may be advisable concerning the further unfolding of electronic marvels in our society. Not everything that is produced by industry is accepted in society. For instance, numerous methods for public transportation have been devised, but only a few have survived in nationwide systems and they show vast differences in many parts of the world. Similarly, the apparent abundance of possibilities in microelectronics will probably be more selectively applied than present developments seem to suggest.

MAIN THEMES: A RECAPITULATION

In this book we have covered four main areas: (1) applications of microelectronics technology, (2) IC product technology and products, (3) supporting research and manpower base, and (4) governmental policies. The following outline may be considered a summary of our main findings, with an American perspective used in each area.

Generally speaking, a vast potential for further progress in microelectronics exists and the technology will find many new applications. On the one hand, the marriage of computer and communication technologies will open up novel perspectives of the technical infrastructure of industrial societies. Related techniques can find broad applications in factories and offices. Especially in offices, new concepts will be developed about automation and its role in the organization. On the other hand, consumer-oriented electronics will also be in a continuous process of change. Not only will new, better, and more versatile versions of present functions become available but new ones will find a place in the home and the individual's life. In part this will be in response to the possibilities engendered by newly developed infrastructures, but new and independent concepts for equipment will also be pursued.

The American industry has a very strong position in most of these new areas of application for this mostly digital technology. Though its performance has been weak in the field of consumer electronics, mistakenly not considered an area for innovation, the U.S. industry has continued its leading role in computers and telecommunications over many years. However, the Japanese industry poses a widely recognized threat. Its telecommunications industry is highly competitive worldwide and its computer industry is expanding swiftly. After their success in consumer electronics, Japanese industrialists will view the U.S. as the major future battlefield in these professional markets.

The underlying integrated-circuit technology will allow industry to continue manufacturing products with a growing capability at a decreased cost. The rate at which this occurs will probably slow down somewhat because the necessary technological advances will become very difficult and need extremely refined production methods. Thus each new step forward is costly and time-consuming. However, physics does not seem to have barriers in sight for the industrial applications envisaged in the next decade.

At the present time, the main practical barriers seem to lie in our inability to master the complexity offered by IC technology. Systems architecture and circuit design are bottlenecks. A somewhat related problem is the cost of producing software that is often associated with digital electronics. Here also, techniques to manage software development have been lagging with respect to the possibilities created by present-day hardware.

Again, the American industrial position in IC's is very strong. Supported by an innovative industry that supplies materials and equipment, the merchant houses have developed products and process technologies that have had no counterpart in the world. Only recently, however, Japanese industry has captured a large share of the market for some advanced circuits. This development has a direct effect on the ability of the U.S. industry to recoup its investments in R&D for novel-process technology. This poses a very serious threat to the U.S. industry's future technological position vis-a-vis that of the Japanese competition, which already seems to be at least on a par with the major U.S merchant houses.

Interestingly, technical change and economic necessities have brought about fundamental changes in the structure of the U.S. industry. A strong trend toward vertical integration has led to the incorporation of many of the IC companies, mostly new ventures of the early 1970's, into larger industrial units. However, in the U.S. there is a perception that large integrated companies are less flexible and innovative than smaller ones. If this proves true in the IC industry it might lead to a further deteriorating position vis-a-vis the Japanese industry, which seems to be able to act and innovate quickly in large industrial conglomerates. In the recent past new venture activity has become apparent, usually aimed at market niches or novel customer services.

The U.S. industry needs a healthy national technology base to guarantee its further expansion. It turns to universities for long-term research that could lead to future products or manufacturing technologies. However, both in industry and academia the impression has been growing that microelectronics research of top quality on subjects of potential industrial interest is limited; the contributions of university research seem to be insufficient to continue sustaining the major role that the IC industry assumes. Various initiatives for improvement, often based on cooperation between university and industry, have been taken. These plans and projects look promising, but they often rely on a narrow financial base and must make use of a highly restricted reservoir of qualified personnel.

The contributions through the federal government to the technology base are scattered (National Science Foundation), limited (National Bureau of Standards), or specialized (Very High Speed Integrated Circuit program). They do not provide a competitive alternative for the Japanese and some of the European governments' efforts.

The skilled manpower available for the industry is marginal in numbers, though well trained. Fears have been expressed that a shortage of engineers in several categories like electronics or informatics will slow down the development of the IC industry. Universities and engineering schools turn out insufficient numbers of graduates. This has been partly due to a diminished interest in an engineering education, but more recently shortages of teachers and laboratory equipment have become limiting factors. Improving this situation is a difficult but necessary task.

The diffusion of microelectronics will have many different but hitherto unforeseeable effects on American society. For instance, though widely publicized, the relations between the introduction of microelectronics and ensuing changes in employment remain unclear. On the one hand, increased productivity will eliminate a number of jobs. On the other hand, new industries and new service companies emerge that create new jobs. Present models are insufficient to give accurate predictions. Thus, it is generally believed in the U.S. that the final reckoning of effects will give a positive result. Many Europeans are less optimistic. In particular, the labor unions wish to exercise influence on the introduction of new technologies to minimize adverse influence on the working environment and job availability.

In summary, the U.S. capability in microelectronics is very strong. However, recent successes of the Japanese IC industry suggest that the U.S. position is not unassailable. Indeed, several structural weaknesses exist, such as the situation of the technology base and the shortage of skilled staff.(3) However, more urgent problems center on the difference in structure of operating costs and profit requirements in the two countries. This allows the Japanese industry more investments in R&D and capital goods and more flexibility in pricesetting. If no corrective actions are taken, this difference may lead to a critical weakening of the position of the U.S. IC industry in the next decade. This, in turn, would bode ill for the American microelectronics industry, which is so critically dependent on a healthy indigenous activity.

The federal government, unlike many other national governments, does not influence the IC industry through direct policies. However, a multitude of measures affect the industry. Some are restrictive, as in the application of antitrust laws. On the other hand, some are supportive to the industry, for instance through defense procurement programs. It seems, however, that the net overall effect of government measures in the U.S. is less supportive to the industry than that in some other countries. Opinions as to the desirability of more support vary. Not infrequently, negative effects accompany government policies. An illustration can be found in the restrictions related to technology transfer imposed on defense-oriented R&D programs, about which much difference in opinion exists. However, though the U.S. has no coherent policy regarding the IC industry, a growing recognition by government agencies of the important role of the industry is apparent.

ACTIONS AND POLICIES

In the course of this book many problem areas have been discussed which merit extensive, in-depth consideration. In this section let us focus on a few issues where the main theme, the interaction of a driving technology with a complex environment, is particularly relevant. Keep in mind that most of these issues are candidates for further analysis and review to arrive at suggestions for actions and policies by the public or the private sector, or by both.

- The business sector which produces process equipment for the semiconductor industry is rather fragmented. The increasing development costs of new equipment will probably lead to problems for several firms. This

- obstacle might begin to weaken the overall competitive strength of the industry. A careful monitoring of this minimally researched industrial sector seems advisable.
- Traditionally, technology in the U.S. is considered to be the property of individual companies. Because of the secrecy involved, it is extremely difficult to assess on a national level the actual strength of the industrial technology in a certain area. Quite often, IC technology is not protected by patents. New mechanisms that could facilitate a more open exchange of information while protecting proprietary knowledge might strengthen the nation's technology base. An exploration of possible mechanisms seems appropriate.
- Innovation is commonly assumed to be more prevalent in the activities of small companies than in those of mature, integrated firms. As the IC industry becomes more mature and vertically integrated, the innovative behavior of the companies may change. How do present policies and emerging trends relate to this assumption?
- In general, recent new ventures in the field do not seem to be highly innovative technologically. A study of their innovative behavior and competitive environment compared with those in earlier waves might be useful for insights into the economic value of new ventures.
- A technical, organizational, and economic study of captive industrial activities is needed to identify their real economic value to companies and how this will develop in the future when VLSI technology poses new problems. What will be the next trend, in view of the new customized-IC trends?
- Microelectronics technology is expected to penetrate parts of industry which have not previously been using electronics. Larger companies usually adapt, but smaller companies may have trouble in adopting new technologies, particularly in the face of shortages of skilled manpower. In several European countries institutions have been established to facilitate the transfer of microelectronics technology to medium and small businesses. A review of American needs seems clearly in order.
- It is evident that in the near future many jobs will be affected by the introduction of microelectronics. In contrast to European countries, not much concern has been expressed in the U.S. relating to these effects on employment. A main reason may have been the flexibility and self-confidence of the American work force compared to the more anxious attitude of Europeans. Another factor may be that full employment is not seen as a direct responsibility of the U.S. federal government. Rather, the ideal American approach is to stimu-

late the economy in general so that more or less automatically a sufficient number of jobs will be created to prevent serious unemployment. Now, however, questions are arising as to whether this approach is still adequate when the job market is expected to change dramatically in a rather brief period. Better insight into the potential problems is needed, such as the consequences of reallocating people to other jobs, what retraining programs are necessary, and what types of jobs will be available.

- The available nationwide data regarding the need for and supply of technical manpower in the IC industry seem to be insufficient to address the problem of manpower shortages forecast by the industry. In view of the rapid changes in the technical environment and international competition, the absence of a national forum for at least the study and discussion of such problems seems to be a distinct weakness for the U.S. Better qualitative and quantitative insights are sorely needed.

- Many universities are not able to provide an adequate education in electrical engineering and computer science. The causes for this failure are multiple. Both level and capacity should be studied in relation to the universities' environment. It seems that only a concerted effort on the part of the federal and state governments, universities, and industry can serve to address the present problem.

- In-company training as a means of improving technical skills of personnel is available in varying degrees. An industry-wide review of the practice and the problems might give insights as to whether the existing work force is used optimally.

- The technology base in microelectronics at many universities is not as strong as might be desirable. Certainly in the present stage where industry is maturing, a solid university research effort should take on more basic, long-term projects. This goal requires a greatly increased funding of such research from both public and private sources. A national policy by NSF, funding centers like Cornell, and new initiatives by industry are needed to achieve a strong and coherent program over time.

- Ties between universities and industry are growing stronger. With an increasing financial contribution from industry, the commitments of universities are changing. An investigation of the various forms of interactions and a discussion of the issues affecting the role of the university in such research will be necessary.

- The direct contribution of the federal government to the microelectronics technology base is rather small. With the exception of the NBS and perhaps some defense laboratories, there is no federal effort in technology development; indeed, the establishment of special federal laboratories for this field is probably undesirable in the American context. Nevertheless it might be useful to consider critically which contributions federal laboratories might make to the broad field of microelectronics, including materials, equipment engineering, design methodology, software, and test equipment. In this respect the effectiveness of technology transfer between federal laboratories and industry should receive careful attention.

- Several comparative studies of the Japanese and American industrial practices are being conducted. A further investigation of how the "fair trade" concepts might be applied are needed. This could provide better evaluative measures for situations that are perceived to be unfair and a rational foundation for necessary remedies.
- The export of technology is now limited only by regulations related to national security. A further restriction of the technology flow abroad, advocated by some as a means of strengthening the national economic position, poses some very intricate problems. A study of the effects on the overall U.S. technical base both in industry and universities and international transfer mechanisms may provide better insights into the extent of the impacts involved.
- The effects of tax-credit mechanisms for capital investment, industrial R&D, and sponsored university R&D may be beneficial to this high-technology industry. An economic study seems appropriate for understanding the advantages and disadvantages of such measures.
- The "microelectronics revolution" brings opportunities and poses threats to the newly developing countries and the underdeveloped countries. Changes in the world economy are likely and will affect U.S. foreign relations. A better insight into these relations is desirable.

FINAL REMARKS

We have to draw attention to the fact that, though microelectronics has become a truly international affair, the options for actions and policies of the U.S. private and public sectors are limited by social and economic traditions, concepts of government responsibility, and other considerations.

The private sector in the U.S. operates in a relatively open system with minimal government interference in comparison with many other countries. This has obvious advantages, but it limits for instance the options for industry cooperation or government protection and support. Many areas of U.S. business, including a major part of the microelectronics industry, are very reluctant to suggest or accept government interference. They basically advocate a hands-off attitude with a minimal tax burden and few regulations as the best way to promote industrial growth. Thus, Japanese or European models for cooperative actions or public policies are not applicable in the American context.

In discussing the American system with respect to public policy, it is worthwhile to keep in mind that other nations have a quite different concept of the responsibilities of a central government. Most European governments are much more closely involved in explicit policies that will support employment and export. A good example of the rationale for such policies is provided by the study by Nora and Minc.(1) More recently, the French government has set up a World Center for the Social Uses of Microelectronics. In Japan, the Ministry of International Trade and Industry recently published its "Vision on the MITI-Policies in 1980's."(2) This document outlines broad proposals for industrial and trade policies, based on a statement of general national goals which again emphasize employment and export trade.

Such European and Japanese concepts go beyond what the American government and private sector consider advisable. Nevertheless, in a situation where world economic competition is increasing and is influencing the U.S. more directly, it may be useful to reconsider what aspects of foreign models might be appropriate.

Microelectronics as a unique example of technical change will continue to influence our world in countless ways. Rapid changes will occur, many issues will require attention, and responses will be formulated. Actions and policies in the past have often been piecemeal reactions to unexpected events rather than strategic approaches that are coordinated and broadly implemented. Though problems exist, exceptional opportunities are present. The capacity to maintain leadership in an extremely complex environment is the real issue. This will require bold action and wise policies that cope with the issues at hand.

REFERENCES

1. Nora, S., and Minc, A., L'Informatisation de la Société (Paris: La Documentation Française, 1978).

2. "The Vision of MITI-Policies in the 1980's," Summary, MITI Information Office, NR-226 (80-7), 1980.

3. Noyce, R.N., "Competition and Cooperation - a Prescription for the Eighties," Research Management, March 1982, p. 13.

Index

Active component, xiii, 5, 46
Actuator, xiii, 25, 66
A/D converter, xiii
ADA, xiii, 18, 68, 69
Advanced Micro Devices, 82, 83, 89
Algorithm, xiii, 31
American Electronics Association (AEA), 151
Analog, xiii, 59
Analog IC's, 71
Application software, xiii
Arpanet, xiii, 132
Artificial intelligence, 19, 114
Assembly language, 67
ATT, 23, 80, 175

Backward integration, 118
Bardeen, John, 46
BASIC, xiii, 67
Bell Laboratories, 3, 18, 46, 92, 97, 147
Berkeley, University of California, 131, 137-138
Bipolar device, xiii, 44, 48, 49, 63, 64, 69, 80

Bit, xiv, 9
BMFT, 123, 159-160
Brattain, W.H., 46
Bubble memory, xiv, 64, 75, 87, 92
Burroughs, 89
Bus, xiv, 32, 35, 66

CAD, xiv, 38, 39, 53, 56, 71, 114
CAM, xiv, 39
California Institute of Technology, 132
Capital investment, 13, 116
Captive industry, 89
CCD, xiv, 74
CDC, 89, 133
Chip, xiv, 55
CIF, xiv, 112, 132
CMOS, xiv, 50, 64, 70, 71, 72, 77, 109
Cobol, xiv, 18
Codec, xiv, 22, 117
COGENT, 173
Commerce Department, 173-174
Computers, 16, 63, 65
Consumer electronics, 27, 40
Control systems, 23, 24, 25

195

D/A converter, xv, 21, 73
D/A conversion, 73
Data general, 89
DEC, 89
Department of Defense, 39, 144-145, 177-178
DES, xv, 36
Design architecture, 6
Die, xv, 55
Digital, xv, 28, 59
Diode, xv, 5
Distributed logic, 16
DMOS, xv, 25, 50
Doping, xv, 46, 47, 57
Dynamic RAM, xv

E-beam machine, xv, 56, 96, 97, 108
ECL, xv, 50, 64, 69
EDP, xv, 16
E PROM, xvi, 61, 62,
EE PROM, xv, 61, 62, 64
EFCIS, 122
Europe, 81, 97, 98
European market, 77
Eurotechnique, 122

Factory automation, 37, 38
Fairchild Camera and Instruments, 82, 84, 86, 88, 96
Federal government, 5, 141, 171-172
FET, xvi, 72
Fibers (optical), xvi, 19, 20, 22, 39
FIFO register, 63
Firmware, xvi, 69
Floppy disc, xvi, 16, 33, 66
Fortran, xvi, 18, 67
Forward integration, 117
Forward pricing, 12, 84
France, 121-122
Fujitsu, 102, 103, 124

Gallium arsenide, xvi, 17, 74, 125
Gate, xvi
Gate arrays, xvi, 111
GATT, xvi, 172
General Electric, 90, 93, 95, 134, 136, 147
General Motors, 90
Government support strategies, 120
GTE, 89

Hall sensor, xvi, 93
Hardware, xvi, 19, 34
Harris, 122
Hewlett-Packard, 80, 90, 94, 108, 147
Hitachi, 87, 102, 124
Honeywell, 89, 90, 93, 133, 147

IBM, 17, 23, 55, 58, 68, 76, 80, 89, 91, 97, 108, 109, 147, 161, 175
IC, xvii, 2, 3, 6, 8, 15, 40, 44, 46, 48, 50, 51, 53, 59, 66, 70, 72, 77, 79
IIL, xvii, 50, 70
In-circuit emulation, 67
Industrial R&D, 147
Information society, 8, 160
Inmos, 123
Intel, 65, 68, 69, 80, 82, 83, 86, 87, 118
Interface, xvii, 66, 73
Intersil, 84
I/O, xvii, 66
Ion implantation, xvii, 57, 96
ISL, xvii, 70
ITAR, xvii, 145, 175
ITEL, 87
ITT, 100

Japan, 4, 38, 39, 40, 55, 80, 81, 124-126, 154

INDEX

Japanese industry, 100
Japanese market, 77
Josephson junction, xvii, 17, 76, 92, 125
Justice Department, 174-175

Kilby, I.S., 48

Learning curve, xvii, 12, 68, 84
Light sensitive devices, 75
Lithography, xviii, 56
LSI, xviii, 6, 48

Mainframe, xviii, 17, 35
Market pull, 3
Masks, xviii, 51, 53, 54
Matra, 122
Matsushita, 102, 103
Mead and Conway, 113
Mead-Conway methodology, 115
Memory, xviii, 9, 60
Merchant house, xviii, 7, 22, 70, 80, 152
Microelectronics Center of North Carolina, 135-136, 138
Microprocessor, xviii, 29, 31, 65, 66, 67, 68, 117
Minicomputer, xviii, 67, 68
Minnesota, University of, 133-134, 138
MIT, 134-135, 137
MITI, xviii, 101, 172
Mitsubishi, 102, 103, 124
MNOS, 62
Modem, xviii, 28
MOS, xviii, 8, 12, 13, 44, 48, 59, 64, 70, 72, 77, 80, 81
Mostek, 80, 82, 83, 84, 88
Motorola, 71, 80, 82, 87, 96, 122

Moore's law, xviii, 9, 10, 108, 110
MSI, xix, 6, 48

National Institutes of Health, xix, 143-144
National Science Foundation, xix, 141-142, 175-177
National Semiconductor, 82, 86, 87, 122
NBS, xix, 145-147
NCR, 89
NEC, 102, 124, 125
New ventures, 4, 79, 119
nMOS, xix, 9, 14, 15, 46, 50, 62, 63, 70
Non-volatile memory, xix, 75
Noyce, R.N., 48, 130
n-silicon, xix, 46

OECD, xix
Office automation, 40
Office equipment, 35
Operational amplifier, xix, 72
Opto-electronics device, xix, 76

PABX, xix, 33
Packaging, 57
PASCAL, xix, 18, 68, 69
Passivation layer, xix, 53
Passive components, xix, 5, 46
Personal computer, 17
Philips, 97, 99, 122
Piezo-electricity, xix, 24
Plan circuits intégrés, 122
pMOS, 50, 70
Prestel, 23, 29
Printed circuit board, xx, 6, 58, 66
Process technology, 57

Programmable logic array, xx, 113
PROM, xx, 61, 62, 64, 65
p-silicon, xx
PTT, xx, 23, 35, 39, 161, 171

RAM, xx, 9, 10, 11, 12, 13, 16, 62, 63, 64, 66
RCA, 90
Resist, xx, 51
Reticle, xx, 54
Robotics, 27, 38, 39
ROM, xx, 60, 65, 69
RTC, 122

Saint-Gobain, 122
Satellite(s), 39
SAW device, xx, 76
Schockley, W.B., 46
Schottky effect, xxi, 50
Schottky-TTL, 69
Second sourcing, xx, 147
Semiconductor Research Cooperative, 180
SESCOSEM, 122
Sensor, xxi, 24
SIA, xxi, 82, 178-180
Siemens, 97, 98, 99
Signetics, 71, 82, 84, 89
Silicon foundries, 71
Silicon valley, xxi, 82, 84
SLIC, xxi, 22, 117
Software, xxi, 17, 18, 19, 24, 36, 67, 68, 69, 114
SOS, xxi, 74
Sperry-Univac, 89, 133
SSI, 6, 48
Stanford University, 136-137
State Department, 174-175
State policies, 180
Static RAM, xxi

Support industry, 96
Systems house, xxi, 117

TDM, xxii
Technology base, 129
Technology push, 2, 3
Tektronix, 90, 96
Telecommunications, 19
Telematics, 160-161
Teletext, xxi, 23, 28
Testing, 57, 110
Texas Instruments, 71, 80, 82, 86, 96, 118
TFT, xxii, 76
Thick film circuit, xxii, 58
Thomson, 122
Toshiba, 102, 103, 124, 125
Transistor, xxii, 3, 5, 6, 46, 48, 61
Transistor-Transistor Logic (TTL), xxii, 50, 69

Uncommitted logic array, xxii, 70
Unemployment, 159-165
United Kingdom, 123-124

VHSIC Program, xxii, 86, 94, 144-145, 177
Viewdata, xxii, 23, 30
VLSI, xxii, 6, 48, 56, 58, 64, 70, 79, 102, 108, 124, 134, 153
VLSI design, 110, 111, 113, 114, 123
VMOS, xxii, 25

Wafer, xxii, 51, 55, 56
West Germany, 123
Western Electric, 89, 92

X-ray lithography, xxii, 108

ZMOS, xxii, 25

About the Authors

NICO HAZEWINDUS obtained his Ph.D. in 1964 from the Technological University of Delft in The Netherlands with a thesis in the field of low-energy nuclear physics. He then joined the Philips Research Laboratories, where he worked on the construction of linear electron accelerators and cyclotrons, ion optics, and electronic equipment for home-based learning. After spending several years in the international coordination of the Philips research organization, he assumed the position of staff member for the coordination of the company's product development activities. In 1981 Dr. Hazewindus spent a half year as a Visiting Fellow at the Center for Science and Technology Policy, New York University, where he conducted the research for this book.

JOHN TOOKER, who collaborated with him, is a student at the Graduate School of Business Administration, New York University. He is currently completing an MBA program with a major in Finance. Prior to this program, he was a student at the Graduate School of Public Administration at New York University and received an MPA degree. His interests include the social and economic consequences resulting from the application of new information-processing technologies, investment in high-technology industries, and government-industry interaction in the microelectronics industry.